AF389386

Recent Advances in Clinical Nutrition in Stroke Rehabilitation

Recent Advances in Clinical Nutrition in Stroke Rehabilitation

Editor

Yoshihiro Yoshimura

MDPI • Basel • Beijing • Wuhan • Barcelona • Belgrade • Manchester • Tokyo • Cluj • Tianjin

Editor
Yoshihiro Yoshimura
Center for Sarcopenia and
Malnutrition Research
Kumamoto Rehabilitation
Hospital
Kumamoto
Japan

Editorial Office
MDPI
St. Alban-Anlage 66
4052 Basel, Switzerland

This is a reprint of articles from the Special Issue published online in the open access journal *Nutrients* (ISSN 2072-6643) (available at: www.mdpi.com/journal/nutrients/special_issues/nutrition_stroke_rehabilitation).

For citation purposes, cite each article independently as indicated on the article page online and as indicated below:

LastName, A.A.; LastName, B.B.; LastName, C.C. Article Title. *Journal Name* **Year**, *Volume Number*, Page Range.

ISBN 978-3-0365-4244-7 (Hbk)
ISBN 978-3-0365-4243-0 (PDF)

Contents

Editorial

Recent Advances in Clinical Nutrition in Stroke Rehabilitation

Yoshihiro Yoshimura 

Center for Sarcopenia and Malnutrition Research, Kumamoto Rehabilitation Hospital, 760 Magate, Kikuyo-Town, Kikuchi-County, Kumamoto 869-1106, Japan; hanley.belfus@gmail.com; Tel.: +81-96-232-3111; Fax: +81-96-232-3119

Citation: Yoshimura, Y. Recent Advances in Clinical Nutrition in Stroke Rehabilitation. *Nutrients* **2022**, *14*, 1130. https://doi.org/10.3390/nu14061130

Received: 14 February 2022
Accepted: 28 February 2022
Published: 8 March 2022

Publisher's Note: MDPI stays neutral with regard to jurisdictional claims in published maps and institutional affiliations.

Stroke is a common cause of death and disability worldwide. Malnutrition is prevalent in stroke rehabilitation patients, and has serious negative effects on outcomes. In addition, there is growing interest in new concepts related to malnutrition, such as sarcopenia, frailty, cachexia, chronic inflammation, dysphagia, and oral problems, all of which contribute to a poor prognosis. Therefore, it is necessary to assess nutritional status early and, if needed, provide appropriate nutritional intervention to improve patient outcomes. A multidisciplinary approach is strongly recommended in this setting; as such, high-quality clinical evidence regarding clinical nutrition in stroke rehabilitation is needed.

This Special Issue updates our knowledge of clinical nutrition for stroke patients and includes interesting studies on topics including nutrition and weight management in the early stages of stroke, the relationship between frailty and improved physical function, weight gain by providing stored energy, physical activity and diet quality, L-carnitine and cognitive level, and the prediction of stroke prognosis using temporal muscles.

Aggressive nutritional management at the early stages of stroke onset may be effective in improving prognosis. In a retrospective cohort study, Sato et al. showed that high-energy nutritional intake during the first week post-stroke was associated with high rates of discharge from the hospital to home [1]. Additionally, Kishimoto et al. showed that weight maintenance or gain in post-stroke patients during the early phases of convalescence rehabilitation is independently associated with improvements in physical function [2]. Furthermore, Yoshimura et al. conducted a retrospective cohort study of underweight patients aged $\geq$70 years with a body mass index of less than 20.0 kg/m^2 undergoing convalescent rehabilitation after stroke. The study found that providing stored energy contributed to weight gain and increased skeletal muscle mass [3], and that it took approximately 9600 kcal of energy to gain 1 kg of body weight in underweight patients. These findings emphasize the importance of not only exercise therapy [4] and correction of polypharmacy [5], but also of aggressive nutritional support at the early stages to improve prognosis post-stroke.

Physical activity is also important in the rehabilitation of stroke patients. Nguyen et al. showed that physical activity and diet quality significantly modified the negative impacts of comorbidity on disability in stroke patients [6]. Comorbidities in stroke patients are strongly associated with poor prognosis, death, increased levels of disability, and worse functional outcomes post-stroke. Therefore, it is important to assess comorbidities early and increase physical activity and diet quality for appropriate treatment and rehabilitation. Furthermore, Nozoe et al. showed that pre-stroke frailty was associated with declines in physical function several months post-stroke [7]. These findings indicate that preventing frailty reduces functional disability, and that maintaining high physical function is associated with a better post-stroke quality of life.

Another interesting finding is that L-carnitine may serve a neuroprotective role against white matter microstructural damage and cognitive impairment in hemodialysis patients, according to Ueno et al. [8]. Long-term administration of carnitine may ameliorate damage to white matter microstructure by suppressing neuroinflammation and improving the executive function and attention associated with the protection of several candidate fiber tracts. Long-term administration of carnitine may be a novel treatment for vascular dementia.

However, as this study is based on hemodialysis patients, further studies are needed to validate the neuroprotective effects of L-carnitine in post-stroke patients.

Katsuki et al. outlined reports on temporal muscle thickness (TMT) and stroke. TMT is associated with nutritional status and risk of sarcopenia after stroke. It is also a useful prognostic marker for dysphagia in patients with subarachnoid and cerebral hemorrhage. In recent years, there has been a rapid increase in the number of reports on TMT and stroke, as TMT is considered one of the most important clinical factors [9]. Sarcopenia and malnutrition are frequently observed in stroke patients and are associated with impaired rehabilitation outcomes. Instruments such as bioelectrical impedance analysis and dual energy X-ray absorptiometry are required to evaluate skeletal muscle mass in diagnosing sarcopenia. However, if TMT can be quantified simply by echo, it may be widely applied in daily stroke rehabilitation clinical practice.

The current Special Issue presents recent advances in clinical nutrition in stroke rehabilitation, highlighting the importance of nutritional management and physical activity in improving functional outcomes after stroke. The advances shown are of great interest from a clinical perspective, with a growing number of stroke patients around the world, and may act as the basis for future research.

Funding: This research received no external funding.

Institutional Review Board Statement: Not applicable.

Informed Consent Statement: Not applicable.

Data Availability Statement: Not applicable.

Conflicts of Interest: The author declares no conflict of interest.

References

1. Sato, Y.; Yoshimura, Y.; Abe, T. Nutrition in the First Week after Stroke Is Associated with Discharge to Home. *Nutrients* **2021**, *13*, 943. [CrossRef] [PubMed]
2. Kishimoto, H.; Nemoto, Y.; Maezawa, T.; Takahashi, K.; Koseki, K.; Ishibashi, K.; Tanamachi, H.; Kobayashi, N.; Kohno, Y. Weight Change during the Early Phase of Convalescent Rehabilitation after Stroke as a Predictor of Functional Recovery: A Retrospective Cohort Study. *Nutrients* **2022**, *14*, 264. [CrossRef] [PubMed]
3. Yoshimura, Y.; Wakabayashi, H.; Momosaki, R.; Nagano, F.; Bise, T.; Shimazu, S.; Shiraishi, A. Stored Energy Increases Body Weight and Skeletal Muscle Mass in Older, Underweight Patients after Stroke. *Nutrients* **2021**, *13*, 3274. [CrossRef] [PubMed]
4. Yoshimura, Y.; Wakabayashi, H.; Nagano, F.; Bise, T.; Shimazu, S.; Shiraishi, A.; Kido, Y.; Matsumoto, A. Chair-Stand Exercise Improves Sarcopenia in Rehabilitation Patients after Stroke. *Nutrients* **2022**, *14*, 461. [CrossRef]
5. Matsumoto, A.; Yoshimura, Y.; Wakabayashi, H.; Kose, E.; Nagano, F.; Bise, T.; Kido, Y.; Shimazu, S.; Shiraishi, A. Deprescribing Leads to Improved Energy Intake among Hospitalized Older Sarcopenic Adults with Polypharmacy after Stroke. *Nutrients* **2022**, *14*, 443. [CrossRef]
6. Nguyen, L.T.K.; Do, B.N.; Vu, D.N.; Pham, K.M.; Vu, M.T.; Nguyen, H.C.; Tran, T.V.; Le, H.P.; Nguyen, T.T.P.; Nguyen, Q.M.; et al. Physical Activity and Diet Quality Modify the Association between Comorbidity and Disability among Stroke Patients. *Nutrients* **2021**, *13*, 1641. [CrossRef] [PubMed]
7. Nozoe, M.; Kubo, H.; Kanai, M.; Yamamoto, M. Relationships between Pre-Stroke SARC-F Scores, Disability, and Risk of Malnutrition and Functional Outcomes after Stroke—A Prospective Cohort Study. *Nutrients* **2021**, *13*, 3586. [CrossRef] [PubMed]
8. Ueno, Y.; Saito, A.; Nakata, J.; Kamagata, K.; Taniguchi, D.; Motoi, Y.; Io, H.; Andica, C.; Shindo, A.; Shiina, K.; et al. Possible Neuroprotective Effects of l-Carnitine on White-Matter Microstructural Damage and Cognitive Decline in Hemodialysis Patients. *Nutrients* **2021**, *13*, 1292. [CrossRef] [PubMed]
9. Katsuki, M.; Kakizawa, Y.; Nishikawa, A.; Yamamoto, Y.; Uchiyama, T.; Agata, M.; Wada, N.; Kawamura, S.; Koh, A. Temporal Muscle and Stroke—A Narrative Review on Current Meaning and Clinical Applications of Temporal Muscle Thickness, Area, and Volume. *Nutrients* **2022**, *14*, 687. [CrossRef]

Review

Temporal Muscle and Stroke—A Narrative Review on Current Meaning and Clinical Applications of Temporal Muscle Thickness, Area, and Volume

Masahito Katsuki [1,2,†] , Yukinari Kakizawa [1,*,†], Akihiro Nishikawa [1], Yasunaga Yamamoto [1], Toshiya Uchiyama [1], Masahiro Agata [1], Naomichi Wada [1], Shin Kawamura [2] and Akihito Koh [2]

[1] Department of Neurosurgery, Suwa Red Cross Hospital, Suwa 392-8510, Nagano, Japan; ktk1122nigt@gmail.com (M.K.); aki.west@gmail.com (A.N.); yamamotoyasunaga@gmail.com (Y.Y.); u_tosh@yahoo.co.jp (T.U.); massa_chiba_3@yahioo.co.jp (M.A.); yosouuemon@gmail.com (N.W.)

[2] Department of Neurosurgery, Itoigawa General Hospital, Itoigawa 941-0006, Niigata, Japan; massa_chiba_1@yahoo.co.jp (S.K.); massa_chiba_2@yahoo.co.jp (A.K.)

* Correspondence: ykakizawajp@yahoo.co.jp

† These authors contributed equally to this work.

Abstract: Background: Evaluating muscle mass and function among stroke patients is important. However, evaluating muscle volume and function is not easy due to the disturbances of consciousness and paresis. Temporal muscle thickness (TMT) has been introduced as a novel surrogate marker for muscle mass, function, and nutritional status. We herein performed a narrative literature review on temporal muscle and stroke to understand the current meaning of TMT in clinical stroke practice. Methods: The search was performed in PubMed, last updated in October 2021. Reports on temporal muscle morphomics and stroke-related diseases or clinical entities were collected. Results: Four studies reported on TMT and subarachnoid hemorrhage, two studies on intracerebral hemorrhage, two studies on ischemic stroke, two studies on standard TMT values, and two studies on nutritional status. TMT was reported as a prognostic factor for several diseases, a surrogate marker for skeletal muscle mass, and an indicator of nutritional status. Computed tomography, magnetic resonance imaging, and ultrasonography were used to measure TMT. Conclusions: TMT is gradually being used as a prognostic factor for stroke or a surrogate marker for skeletal muscle mass and nutritional status. The establishment of standard methods to measure TMT and large prospective studies to further investigate the relationship between TMT and diseases are needed.

Keywords: frailty; muscle volume; nutritional status; prognostic factor; sarcopenia; skeletal muscle mass; stroke; temporal muscle thickness

Citation: Katsuki, M.; Kakizawa, Y.; Nishikawa, A.; Yamamoto, Y.; Uchiyama, T.; Agata, M.; Wada, N.; Kawamura, S.; Koh, A. Temporal Muscle and Stroke—A Narrative Review on Current Meaning and Clinical Applications of Temporal Muscle Thickness, Area, and Volume. *Nutrients* **2022**, *14*, 687. https://doi.org/10.3390/nu14030687

Academic Editors: Yoshihiro Yoshimura and Kosaku Kinoshita

Received: 30 November 2021
Accepted: 4 February 2022
Published: 6 February 2022

Publisher's Note: MDPI stays neutral with regard to jurisdictional claims in published maps and institutional affiliations.

1. Introduction

Stroke is a widely known cause of disability [1]. Stroke also increases the risk of skeletal muscle loss [2]—sarcopenia—which contributes to further disability related to stroke [3]. Furthermore, pre-stroke sarcopenia is also associated with poor functional outcomes [4,5]. Therefore, evaluating muscle mass and function among stroke patients is important [6,7], and aggressive nutrition therapy [8–10], deprescribing [11], and rehabilitation [12,13] re applicable for those with stroke, as well as those at high risk for muscle loss.

Measuring skeletal muscle mass and function is an evolving parameter for the clinical evaluation of physiological conditions [14]. The gold standard to evaluate sarcopenia are muscle function tests such as the gait speed test and the grip strength test, according to the European Working Group on Sarcopenia in Older People (EWGSOP), EWGSOP2, and the Asian Working Group for Sarcopenia (AGWS) [15,16]. However, measuring muscle function such as grip strength and gait speed sometimes cannot be performed because stroke patients often have disturbances of consciousness, are sedated, are resting due to surgical treatment, or experience paresis. Therefore, an alternative method to evaluate muscle mass and function is needed.

Recently, temporal muscle thickness (TMT) on computed tomography (CT) images or magnetic resonance images (MRI) has been introduced as a novel surrogate marker with which to measure muscle mass [17], function [18], and nutritional status [19,20]. CT and MRI are routinely performed for stroke patients, and TMT measurement is easier than other methods such as quantitative measurements of the cross-sectional skeletal muscle area at the third lumbar vertebra using CT imaging, which is known to be significantly correlated with whole-body muscle [16]. Therefore, TMT is attractive as an alternative method to evaluate muscle mass and function for stroke patients. The first purpose of the narrative literature review that we present herein is to investigate reports on TMT and stroke. The second purpose is to understand the current meaning of TMT in clinical stroke practice. In addition to TMT [21,22], we also examined temporal muscle area (TMA) [21,23,24] and temporal muscle volume (TMV) [9] as novel TM-related surrogate markers for skeletal muscle mass.

2. Materials and Methods

Studies regarding TMT and stroke were examined. The search was performed in PubMed, last updated in October 2021, using the terms "stroke" OR "intracerebral hemorrhage (ICH)" OR "subarachnoid hemorrhage (SAH)" OR "cerebral infarction" OR "rehabilitation" OR "sarcopenia" OR "frailty" OR "nutrition" AND "temporal muscle thickness". The PubMed search resulted in a total of 73 articles. We systematically read through the abstracts of all original articles available in English. We included studies on the association between stroke and TMT, with a sample size of around 50 cases and appropriate statistical analyses. We also checked through the lists of references to complete our collection of studies. All the authors verified the correct transcription of the data to our manuscript. Finally, we included eight studies related to stroke in our review.

3. Results

Four studies reported the association between TMT and SAH [9,21,23,24], two studies between TMT and ICH [25,26], and two studies between TMT and ischemic stroke [27,28]. The other four studies described the standard TMT values [18,29] and the relationship between TMT and nutritional status [19,20]. The Preferred Reporting Items for Systematic Reviews and Meta-Analyses (PRISMA) 2020 flow diagram [30] for systematic review is shown as Figure 1.

3.1. Temporal Muscle and SAH

Katsuki et al. first reported TMT as a prognostic factor for SAH outcomes in 2019, investigating 49 SAH patients over 75 years of age who were treated by clipping with craniotomy [21]. TMT was measured on CT images on admission, using Aquilion ONE (Canon Medical Systems Corporation, Tochigi, Japan) with $0.5 \times 0.5 \times 1.0$ mm voxels. The slice thickness was reconstructed to 5 mm. The window width was adjusted to 300 Hounsfield units and the window level was adjusted to 20 Hounsfield units. TMT was measured bilaterally perpendicular to the long axis of the temporal muscle at a slice 5 mm above the orbital roof using SYNAPSE V 4.1.5 imaging software (Fujifilm Medical, Tokyo, Japan). Then, the averages of the left and right of the TMTs were used. The method to measure TMT on CT was thereby defined. Katsuki et al. then performed univariate analysis regarding TMT and functional outcome at six months. The study was preliminary, but the study suggested that greater TMT was related to favorable outcomes among elderly SAH.

Katsuki et al. next investigated the relationship between temporal muscle and Hunt and Kosnik grade on admission and functional outcome at six months [23]. They examined 298 all age-group patients, and all patients were treated by endovascular coiling. They revealed that the Hunt and Kosnik grade on admission and functional outcome were related to TMT and TMA. TMA was measured manually by tracing the outline of the temporal muscle on the same CT slice as that used for measuring TMT. Notably, this study suggests that TMT and TMA are related to both the severity of SAH and functional outcome regardless of age, not only for the elderly.

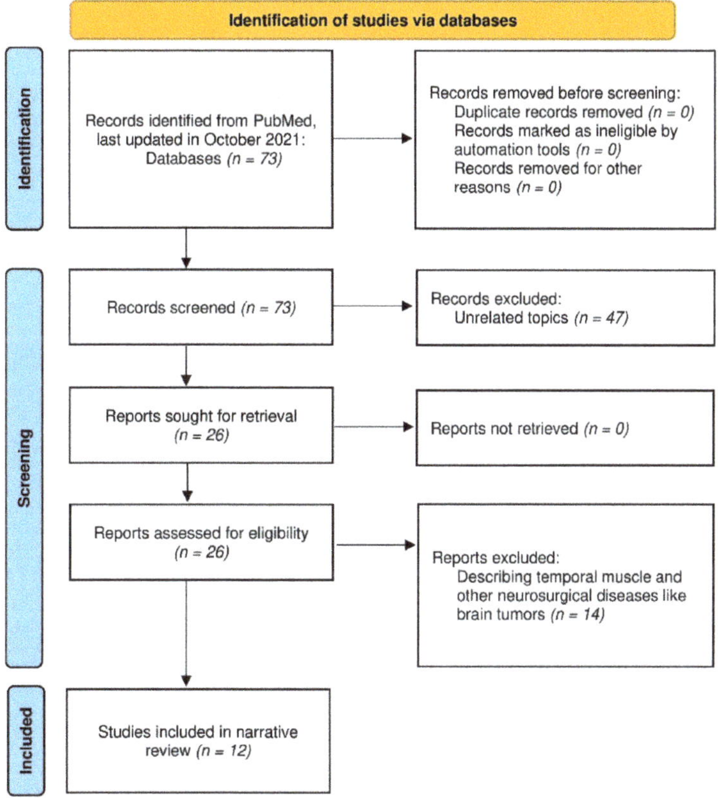

Figure 1. PRISMA 2020 flow diagram for this review.

They then investigated 127 SAH patients under 75 years of age who were treated by clipping [24]. They examined the cut-off values for the functional outcomes. Receiver operating characteristic analysis found that the threshold of TMT was 4.9 mm in women and 6.7 mm in men, and that of TMA was 193 mm^2 in women and 333 mm^2 in men, which were the cut-off values for the functional outcomes among SAH patients under 75 years of age.

Onodera et al. [9] examined TMV using volume rendering software (Ziostation 2 version 2.9.5.1, Ziosoft, Tokyo), because TMT may be less reproducible. They investigated 60 SAH patients and measured TMV on the CT images at admission and two weeks after aneurysm treatment. Patients whose TMV had decreased by $\geq$20% were classified into the "atrophy group," whereas those whose TMV had decreased by <20% were classified into the "maintenance group." Their study showed that the food intake score and the functional outcome were significantly more positive in the TMV maintenance group than the TMV atrophy group. Therefore, this study suggests the importance of early high-protein administration to maintain TMV in the acute term (Table 1).

Table 1. Previous reports on the association between temporal muscle and SAH.

Author	Year	Number of Cases	Abstract
Katsuki [21]	2019	49	High TMT was related to favorable outcomes among elderly SAH.
Katsuki [23]	2020	298	TMT and TMA were related to Hunt and Kosnik grade and functional outcome at six months after endovascular coiling, regardless of age.
Katsuki [24]	2021	127	The threshold of TMT was 4.9 mm in women and 6.7 mm in men, and that of TMA was 193 mm^2 in women and 333 mm^2 in men, which were the cut-off values for the functional outcomes at six months among SAH patients under 75 years of age.
Onodera [9]	2021	60	The food intake score and the functional outcome at discharge were significantly more positive in the TMV maintenance group than the TMV atrophy group after SAH.

Abbreviations: SAH: subarachnoid hemorrhage; TMA: temporal muscle area; TMT: temporal muscle thickness; TMV: temporal muscle volume.

3.2. Temporal Muscle and ICH

Katsuki et al. examined 75 ICH patients treated by endoscopic hematoma removal and investigated the factors related to the functional outcome [25]. They revealed that lower total protein level was related to poor outcomes at six months. In addition, they mentioned TMA as an indicator of nutrition, but TMA itself was not significantly related to the outcome ($p = 0.08$). However, they suggested that low nutritional status, indicated by lower total protein level and low TMA altogether, seemed to be associated with poor outcomes.

Gomes et al. examined 24 post-hemorrhagic stroke patients in the chronic stage and tested bite force and TMT [26]. Maximum molar bite force was verified using a digital dynamometer. TMT was measured using ultrasound images obtained at rest and during maximal voluntary contraction of the masseter and temporalis muscles. The TMT on the unaffected side was larger than on the affected side. This study first focused on the functional and morphological changes in the stomatognathic system after a hemorrhagic stroke. The clinical meaning of these changes was investigated (Table 2).

Table 2. Previous reports on the association between temporal muscle and ICH.

Author	Year	Number of Cases	Abstract
Katsuki [25]	2019	75	Low nutritional status, indicated by low total protein level and low TMA altogether, seemed to be associated with the poor functional outcomes at six months after endoscopic hematoma removal.
Gomes [26]	2021	24	TMT on the unaffected side was greater than on the affected side after a hemorrhagic stroke.

Abbreviations: ICH: intracerebral hemorrhage; TMA: temporal muscle area; TMT: temporal muscle thickness.

3.3. Temporal Muscle and Stroke

Sakai et al. [27] investigated 70 acute cerebral infarction patients' TMT on the T2-weight MR image and functional oral intake scales. They revealed that TMT was a significant explanator of dysphagia severity following acute ischemic stroke, along with age and the National Institute of Health Stroke Scale score. The measuring method of TMT using T2-weighted images was similar to the previous report from Furtner et al. using T1-weighted images [31]. They first reported the association between TMT and ischemic stroke-related dysphagia in the acute term.

Nozoe et al. [28] examined 289 acute elderly stroke patients and investigated TMT on CT images as an indicator of sarcopenia risk and its relationship with the functional outcome at three months. They found that sarcopenia risk was independently associated with TMT in older patients with acute stroke. However, TMT was not independently related to the functional outcome (Table 3).

Table 3. Previous reports on the association between temporal muscle and stroke.

Author	Year	Number of Cases	Abstract
Sakai [27]	2021	70	TMT was a significant explanator of dysphagia severity following acute ischemic stroke.
Nozoe [28]	2021	289	Sarcopenia risk was independently associated with TMT in older patients with acute stroke, but TMT was not independently related to the functional outcome.

Abbreviations: TMT: temporal muscle thickness.

3.4. Standard Values of TMT

Steindl et al. [18] investigated a 624-individual MRI dataset to establish standard reference values of TMT on T1-weighted images. The cohort consisted of two MRI repositories: The Enhanced Nathan Kline Institute-Rockland Sample [32]; and the Designed Database of MR Brain Images of Healthy Volunteers [33]. TMT was measured on isovoxel $(1 \times 1 \times 1 \text{ mm}^3)$ T1-weighted MR images perpendicular to the long axis of the temporal muscle on an axial plane, which was oriented parallel to the anterior commissure-posterior commissure line. They also examined 422 healthy volunteers and 130 cases as a prospective validation cohort and found that TMT and grip strength were correlated. This was the first report to validate the relationship between the TMT and grip strength, namely muscle function, prospectively.

Katsuki et al. [29] investigated a database of 360 Japanese individuals' brain check-ups obtained by MRI. They measured TMT in the same way previously reported in [18] to obtain standard values of TMT among Japanese individuals. They compared their result to Steindl's results to obtain the racial difference, but the background of the participants differed. They did not perform any muscle function test, so further investigation is needed (Table 4).

Table 4. TMT and nutritional status.

Author	Year	Number of Cases	Abstract
Steindl [18]	2020	1175	Standard values of TMT were investigated, and TMT and grip strength were correlated.
Katsuki [29]	2021	360	Standard values of TMT were investigated among Japanese individuals who underwent brain check-ups.

Abbreviations: TMT: temporal muscle thickness.

3.5. TMT and Nutritional Status

Hasegawa et al. [20] investigated 73 elderly individuals to measure their TMT using ultrasonography and nutritional status assessed with anthropometric measurements and laboratory tests. Arm circumference (AC) was measured in the middle of the non-dominant upper arm using a measuring tape. Arm muscle circumference (AMC) was calculated based on the standard procedure using the following formula: AMC (cm) = AC (cm)$-\pi$ x the triceps skinfold thickness (cm) [34]. Calf circumference (CC) was measured with a tape at the maximum girth of the right calf with the leg in a lying position. TMT was strongly correlated with CC and AMC. However, there were no strong correlations with serum protein levels, nor was fat mass evaluated in the triceps skinfold thickness. They also examined the reliability to measure TMT using ultrasonography; the inter-rater reliability was 0.99.

Hasegawa et al. also performed a prospective study [19]. The study aimed to examine whether a change in TMT evaluated by the ultrasonography was directly correlated with energy adequacy, and to determine the cut-off value of a change in TMT to detect energy inadequacy. They investigated 48 bedridden elderly patients and revealed that percentage change in TMT was significantly correlated with energy adequacy. They suggested that the assessment of TMT changes could be helpful for performing better nutritional therapy (Table 5).

Table 5. TMT and nutritional status.

Author	Year	Number of Cases	Abstract
Hasegawa [20]	2019	73	TMT was strongly correlated with CC and ACM. However, there were no strong correlations with serum protein levels, nor was fat mass evaluated in the triceps skinfold thickness.
Hasegawa [19]	2021	48	TMT changes were directly correlated with energy adequacy in bedridden older adults.

Abbreviations: ACM: arm muscle circumference; CC: Calf circumference; TMT: temporal muscle thickness.

4. Discussion

We herein reviewed reports on TMT and stroke. TMT is useful as a prognostic marker for SAH, ICH, and dysphagia after stroke. It also indicates nutritional status and risk of sarcopenia. As the number of reports on TMT and stroke has been increasing rapidly in recent years, we believe that TMT is one of the important factors in clinical practice. In addition to this review, we discussed the TMT measurement method and TMT use in other neurosurgical practices.

4.1. TMT Measurement Method

A standard TMT measurement method has not been established. Old reports used volume rendering software [35,36], and Onodera et al. also used a similar approach [9] to measure TMV, but not TMT. Then, Furtner et al. established TMT measurement using T1-weighted MR images. They measured TMT perpendicular to the long axis of the temporal muscle at the level of the orbital roof [14,17,18,37,38]. This method is widely used, but low accessibility to MRI in routine work is a problem. Sakai et al. [27] used T2-weighted MR images, rather than T1-weighted images. The difference between the T1- and T2-weighted images should be discussed. Katsuki et al.first defined TMT and TMA on CT images [21]. CT is more accessible than MRI, so TMT measurement on CT seems better for routine clinical work. Hasegawa et al. used ultrasonography (M-Turbo; SonoSite, Bothell, WA, USA) to measure TMT at 4 cm from the eyelid and 2 cm above the reference line, which was the orbitomeatal line [20]. Ultrasonography is not so reproducible, but their study reported that TMT measurement by ultrasonography is reliable.

As described above, there are some ways to measure temporal muscle morphomics, including TMT, TMA, and TMV. Easiness and high reproducibility are needed to establish a standard method. Further study on the measurement method is desirable.

4.2. Temporal Muscle in Other Neurosurgical Practice

The first report on the temporal muscle as a prognostic factor in neurosurgical practice evaluated the operative risk in non-syndromic craniosynostosis in 2013 [36]. The authors used volume rendering software to assess the temporal fat pad. Since this report, there have been several papers on the temporalis muscle and prognosis, especially in brain tumors. There are several reports on overall survival and temporal muscle in glioblastoma [38–45], metastatic brain tumor [31,46,47], and primary central nervous system lymphoma [37,48]. As in reports on TMT and stroke, all of these reports used temporal muscle to indicate nutritional status and skeletal muscle mass volume. The greater the temporal muscle, the better the outcome, probably due to better nutritional status and more skeletal muscle mass. Furthermore, deep learning-based quantification of TMA has been reported [49], so it is expected that TMA measuring will be widely performed.

4.3. Limitations

As described above, TMT is now attractive, and many studies have been performed, but some issues should be addressed. First, most of the studies were retrospective, so further prospective study is needed. Second, the sample sizes were small, so studies with large sample sizes are desirable. Third, a standard TMT measurement method has not been established, and several methods can be used, such as MRI, CT, and ultrasonography. A standard approach to measuring TMT is needed. Fourth, the direct mechanism of why large temporal muscle relates to favorable prognosis has not been clarified. The true mechanism between TMT and outcomes should be discussed from several perspectives, such as rehabilitation, nutrition, frailty, deglutition, or basic medicine. Some of the problems may be resolved as TMT measurements are routinely taken, thereby tracking time-course changes.

5. Conclusions

TMT seems to be useful surrogate marker for skeletal muscle volume and function, and is a potential prognostic factor. Research on the association between stroke and TMT is increasing. Further research is needed to establish the usefulness of TMT.

Funding: None.

Institutional Review Board Statement: Not applicable.

Informed Consent Statement: This is a review of literature, so informed consent statement is not applicable.

Data Availability Statement: Not applicable.

Acknowledgments: We thank the hospital staff.

Conflicts of Interest: The authors declare no conflict of interest.

References

1. Toyoda, K.; Sonoda, K.; Sato, S.; Yoshimura, S. The Japan Stroke Databank -Current stroke care in Japan and ideal stroke registries. *Neurol. Ther.* **2018**, *35*, 188–192. (In Japanese) [CrossRef]
2. Ryan, A.S.; Ivey, F.M.; Serra, M.C.; Hartstein, J.; Hafer-Macko, C.E. Sarcopenia and Physical Function in Middle-Aged and Older Stroke Survivors. *Arch. Phys. Med. Rehabilitation* **2016**, *98*, 495–499. [CrossRef]
3. Scherbakov, N.; von Haehling, S.; Anker, S.D.; Dirnagl, U.; Doehner, W. Stroke induced Sarcopenia: Muscle wasting and disability after stroke. *Int. J. Cardiol.* **2013**, *170*, 89–94. [CrossRef]
4. Yoshimura, Y.; Wakabayashi, H.; Bise, T.; Nagano, F.; Shimazu, S.; Shiraishi, A.; Yamaga, M.; Koga, H. Sarcopenia is associated with worse recovery of physical function and dysphagia and a lower rate of home discharge in Japanese hospitalized adults undergoing convalescent rehabilitation. *Nutrition* **2019**, *61*, 111–118. [CrossRef]
5. Nozoe, M.; Kanai, M.; Kubo, H.; Yamamoto, M.; Shimada, S.; Mase, K. Prestroke sarcopenia and functional outcomes in elderly patients who have had an acute stroke: A prospective cohort study. *Nutrition* **2019**, *66*, 44–47. [CrossRef]
6. Nozoe, M.; Kubo, H.; Kanai, M.; Yamamoto, M. Relationships between Pre-Stroke SARC-F Scores, Disability, and Risk of Malnutrition and Functional Outcomes after Stroke—A Prospective Cohort Study. *Nutrients* **2021**, *13*, 3586. [CrossRef]
7. Nakanishi, N.; Okura, K.; Okamura, M.; Nawata, K.; Shinohara, A.; Tanaka, K.; Katayama, S. Measuring and Monitoring Skeletal Muscle Mass after Stroke: A Review of Current Methods and Clinical Applications. *J. Stroke Cerebrovasc. Dis.* **2021**, *30*, 105736. [CrossRef]
8. Yoshimura, Y.; Wakabayashi, H.; Momosaki, R.; Nagano, F.; Bise, T.; Shimazu, S.; Shiraishi, A. Stored Energy Increases Body Weight and Skeletal Muscle Mass in Older, Underweight Patients after Stroke. *Nutrients* **2021**, *13*, 3274. [CrossRef]
9. Onodera, H.; Mogamiya, T.; Matsushima, S.; Sase, T.; Kawaguchi, K.; Nakamura, H.; Sakakibara, Y. High protein intake after subarachnoid hemorrhage improves oral intake and temporal muscle volume. *Clin. Nutr.* **2021**, *40*, 4187–4191. [CrossRef]
10. Onodera, H.; Mogamiya, T.; Mori, M.; Matsushima, S.; Sase, T.; Nakamura, H.; Sakakibara, Y. High protein intake after subarachnoid hemorrhage improves ingestion function and temporal muscle volume. *Clin. Nutr. ESPEN* **2020**, *40*, 595. [CrossRef]
11. Matsumoto, A.; Yoshimura, Y.; Wakabayashi, H.; Kose, E.; Nagano, F.; Bise, T.; Kido, Y.; Shimazu, S.; Shiraishi, A. Deprescribing Leads to Improved Energy Intake among Hospitalized Older Sarcopenic Adults with Polypharmacy after Stroke. *Nutrients* **2022**, *14*, 443. [CrossRef]
12. Yoshimura, Y.; Wakabayashi, H.; Nagano, F.; Bise, T.; Shimazu, S.; Shiraishi, A.; Kido, Y.; Matsumoto, A. Chair-Stand Exercise Improves Sarcopenia in Rehabilitation Patients after Stroke. *Nutrients* **2022**, *14*, 461. [CrossRef]

13. Yoshimura, Y. Treating sarcopenia with a hybrid of aggressive exercise therapy and nutritional therapy: Approaches at Kumamoto Rehabilitation Hospital. *Mon. B Med. Rehabil.* **2018**, *224*, 16–24. (In Japanese)

14. Furtner, J.; Weller, M.; Weber, M.; Gorlia, T.; Nabors, B.; Reardon, D.A.; Tonn, J.-C.; Stupp, R.; Preusser, M. Temporal Muscle Thickness as a Prognostic Marker in Patients with Newly Diagnosed Glioblastoma: Translational Imaging Analysis of the CENTRIC EORTC 26071–22072 and CORE Trials. *Clin. Cancer Res.* **2021**, *28*, 129–136. [CrossRef]

15. Chen, L.-K.; Liu, L.-K.; Woo, J.; Assantachai, P.; Auyeung, T.-W.; Bahyah, K.S.; Chou, M.-Y.; Chen, L.-Y.; Hsu, P.-S.; Krairit, O.; et al. Sarcopenia in Asia: Consensus Report of the Asian Working Group for Sarcopenia. *J. Am. Med. Dir. Assoc.* **2014**, *15*, 95–101. [CrossRef]

16. Cruz-Jentoft, A.J.; Bahat, G.; Bauer, J.; Boirie, Y.; Bruyère, O.; Cederholm, T.; Cooper, C.; Landi, F.; Rolland, Y.; Sayer, A.A.; et al. Sarcopenia: Revised European consensus on definition and diagnosis. *Age Ageing* **2019**, *48*, 16–31. [CrossRef]

17. Leitner, J.; Pelster, S.; Schöpf, V.; Berghoff, A.S.; Woitek, R.; Asenbaum, U.; Nenning, K.-H.; Widhalm, G.; Kiesel, B.; Gatterbauer, B.; et al. High correlation of temporal muscle thickness with lumbar skeletal muscle cross-sectional area in patients with brain metastases. *PLoS ONE* **2018**, *13*, e0207849. [CrossRef]

18. Steindl, A.; Leitner, J.; Schwarz, M.; Nenning, K.-H.; Asenbaum, U.; Mayer, S.; Woitek, R.; Weber, M.; Schöpf, V.; Berghoff, A.S.; et al. Sarcopenia in Neurological Patients: Standard Values for Temporal Muscle Thickness and Muscle Strength Evaluation. *J. Clin. Med.* **2020**, *9*, 1272. [CrossRef]

19. Hasegawa, Y.; Yoshida, M.; Sato, A.; Fujimoto, Y.; Minematsu, T.; Sugama, J.; Sanada, H. A change in temporal muscle thickness is correlated with past energy adequacy in bedridden older adults: A prospective cohort study. *BMC Geriatr.* **2021**, *21*, 1–9. [CrossRef]

20. Hasegawa, Y.; Yoshida, M.; Sato, A.; Fujimoto, Y.; Minematsu, T.; Sugama, J.; Sanada, H. Temporal muscle thickness as a new indicator of nutritional status in older individuals. *Geriatr. Gerontol. Int.* **2019**, *19*, 135–140. [CrossRef]

21. Katsuki, M.; Yamamoto, Y.; Uchiyama, T.; Wada, N.; Kakizawa, Y. Clinical characteristics of aneurysmal subarachnoid hemorrhage in the elderly over 75; would temporal muscle be a potential prognostic factor as an indicator of sarcopenia? *Clin. Neurol. Neurosurg.* **2019**, *186*, 105535. [CrossRef] [PubMed]

22. Gonçalves, R.C.G.; Rabelo, N.N.; Figueiredo, E.G.; Welling, L.C. Oral health and temporal muscle thickness. *Surg. Neurol. Int.* **2021**, *12*, 527. [CrossRef] [PubMed]

23. Katsuki, M.; Suzuki, Y.; Kunitoki, K.; Sato, Y.; Sasaki, K.; Mashiyama, S.; Matsuoka, R.; Allen, E.; Saimaru, H.; Sugawara, R.; et al. Temporal Muscle as an Indicator of Sarcopenia is Independently Associated with Hunt and Kosnik Grade on Admission and the Modified Rankin Scale Score at 6 Months of Patients with Subarachnoid Hemorrhage Treated by Endovascular Coiling. *World Neurosurg.* **2020**, *137*, e526–e534. [CrossRef] [PubMed]

24. Katsuki, M.; Kakizawa, Y.; Nishikawa, A.; Yamamoto, Y.; Uchiyama, T. Temporal muscle thickness and area are an independent prognostic factors in patients aged 75 or younger with aneurysmal subarachnoid hemorrhage treated by clipping. *Surg. Neurol. Int.* **2021**, *12*, 151. [CrossRef]

25. Katsuki, M.; Kakizawa, Y.; Nishikawa, A.; Yamamoto, Y.; Uchiyama, T. Lower total protein and absence of neuronavigation are novel poor prognostic factors of endoscopic hematoma removal for intracerebral hemorrhage. *J. Stroke Cerebrovasc. Dis.* **2020**, *29*, 105050. [CrossRef]

26. Gomes, G.G.C.; Palinkas, M.; da Silva, G.P.; Gonçalves, C.R.; Lopes, R.F.T.; Verri, E.D.; Fabrin, S.C.V.; Fioco, E.M.; Siéssere, S.; Regalo, S.C.H. Bite Force, Thickness, and Thermographic Patterns of Masticatory Muscles Post-Hemorrhagic Stroke. *J. Stroke Cerebrovasc. Dis.* **2021**, *31*, 106173. [CrossRef]

27. Sakai, K.; Katayama, M.; Nakajima, J.; Inoue, S.; Koizumi, K.; Okada, S.; Suga, S.; Nomura, T.; Matsuura, N. Temporal muscle thickness is associated with the severity of dysphagia in patients with acute stroke. *Arch. Gerontol. Geriatr.* **2021**, *96*, 104439. [CrossRef]

28. Nozoe, M.; Kubo, H.; Kanai, M.; Yamamoto, M.; Okakita, M.; Suzuki, H.; Shimada, S.; Mase, K. Reliability and validity of measuring temporal muscle thickness as the evaluation of sarcopenia risk and the relationship with functional outcome in older patients with acute stroke. *Clin. Neurol. Neurosurg.* **2021**, *201*, 106444. [CrossRef]

29. Katsuki, M.; Narita, N.; Sasaki, K.; Sato, Y.; Suzuki, Y.; Mashiyama, S.; Tominaga, T. Standard values for temporal muscle thickness in the Japanese population who undergo brain check-up by magnetic resonance imaging. *Surg. Neurol. Int.* **2021**, *12*, 67. [CrossRef]

30. Page, M.J.; McKenzie, J.E.; Bossuyt, P.M.; Boutron, I.; Hoffmann, T.C.; Mulrow, C.D.; Shamseer, L.; Tetzlaff, J.M.; Akl, E.A.; Brennan, S.E.; et al. The PRISMA 2020 statement: An updated guideline for reporting systematic reviews. *BMJ* **2021**, *372*, n71. [CrossRef]

31. Furtner, J.; Berghoff, A.S.; Albtoush, O.M.; Woitek, R.; Asenbaum, U.; Prayer, D.; Widhalm, G.; Gatterbauer, B.; Dieckmann, K.; Birner, P.; et al. Survival prediction using temporal muscle thickness measurements on cranial magnetic resonance images in patients with newly diagnosed brain metastases. *Eur. Radiol.* **2017**, *27*, 3167–3173. [CrossRef] [PubMed]

32. Nooner, K.B.; Colcombe, S.J.; Tobe, R.H.; Mennes, M.; Benedict, M.M.; Moreno, A.L.; Panek, L.J.; Brown, S.; Zavitz, S.T.; Li, Q.; et al. The NKI-Rockland Sample: A Model for Accelerating the Pace of Discovery Science in Psychiatry. *Front. Behav. Neurosci.* **2012**, *6*, 152. [CrossRef]

33. Bullitt, E.; Zeng, D.; Gerig, G.; Aylward, S.; Joshi, S.; Smith, J.K.; Lin, W.; Ewend, M.G. Vessel Tortuosity and Brain Tumor Malignancy: A Blinded Study1. *Acad. Radiol.* **2005**, *12*, 1232–1240. [CrossRef]

34. Moriwaki, E.-I.; Enomoto, H.; Saito, M.; Hara, N.; Nishikawa, H.; Nishimura, T.; Iwata, Y.; Iijima, H.; Nishiguchi, S. The Anthropometric Assessment with the Bioimpedance Method Is Associated with the Prognosis of Cirrhotic Patients. *Vivo* **2020**, *34*, 687–693. [CrossRef]

35. Ranganathan, K.; Terjimanian, M.; Lisiecki, J.; Rinkinen, J.; Mukkamala, A.; Brownley, C.; Buchman, S.R.; Wang, S.C.; Levi, B. Temporalis muscle morphomics: The psoas of the craniofacial skeleton. *J. Surg. Res.* **2014**, *186*, 246–252. [CrossRef]

36. Rinkinen, J.; Zhang, P.; Wang, L.; Enchakalody, B.; Terjimanian, M.; Holcomb, S.; Wang, S.C.; Buchman, S.R.; Levi, B. Novel Temporalis Muscle and Fat Pad Morphomic Analyses Aids Preoperative Risk Evaluation and Outcome Assessment in Nonsyndromic Craniosynostosis. *J. Craniofacial Surg.* **2013**, *24*, 250–255. [CrossRef] [PubMed]

37. Furtner, J.; Nenning, K.-H.; Roetzer, T.; Gesperger, J.; Seebrecht, L.; Weber, M.; Grams, A.; Leber, S.; Marhold, F.; Sherif, C.; et al. Evaluation of the Temporal Muscle Thickness as an Independent Prognostic Biomarker in Patients with Primary Central Nervous System Lymphoma. *Cancers* **2021**, *13*, 566. [CrossRef] [PubMed]

38. Furtner, J.; Genbrugge, E.; Gorlia, T.; Bendszus, M.; Nowosielski, M.; Golfinopoulos, V.; Weller, M.; Bent, M.J.V.D.; Wick, W.; Preusser, M. Temporal muscle thickness is an independent prognostic marker in patients with progressive glioblastoma: Translational imaging analysis of the EORTC 26101 trial. *Neuro-Oncol.* **2019**, *21*, 1587–1594. [CrossRef] [PubMed]

39. Hsieh, K.; Hwang, M.; Estevez-Inoa, G.; Saraf, A.; Spina, C.; Smith, D.; Wu, C.; Wang, T. Temporalis Muscle Width as a Measure of Sarcopenia Independently Predicts Overall Survival in Patients with Newly Diagnosed Glioblastoma. *Int. J. Radiat. Oncol.* **2018**, *102*, e225. [CrossRef]

40. An, G.; Ahn, S.; Park, J.-S.; Jeun, S.S.; Kil Hong, Y. Association between temporal muscle thickness and clinical outcomes in patients with newly diagnosed glioblastoma. *J. Cancer Res. Clin. Oncol.* **2020**, *147*, 901–909. [CrossRef]

41. Cinkir, H.Y.; Er, H.C. Is temporal muscle thickness a survival predictor in newly diagnosed glioblastoma multiforme? *Asia-Pacific J. Clin. Oncol.* **2020**, *16*, e223–e227. [CrossRef]

42. Muglia, R.; Simonelli, M.; Pessina, F.; Morenghi, E.; Navarria, P.; Persico, P.; Lorenzi, E.; Dipasquale, A.; Grimaldi, M.; Scorsetti, M.; et al. Prognostic relevance of temporal muscle thickness as a marker of sarcopenia in patients with glioblastoma at diagnosis. *Eur. Radiol.* **2020**, *31*, 4079–4086. [CrossRef] [PubMed]

43. Liu, F.; Xing, D.; Zha, Y.; Wang, L.; Dong, W.; Li, L.; Gong, W.; Hu, L. Predictive Value of Temporal Muscle Thickness Measurements on Cranial Magnetic Resonance Images in the Prognosis of Patients with Primary Glioblastoma. *Front. Neurol.* **2020**, *11*, 523292. [CrossRef] [PubMed]

44. Guven, D.C.; Aksun, M.S.; Cakir, I.Y.; Kilickap, S.; Kertmen, N. The association of BMI and sarcopenia with survival in patients with glioblastoma multiforme. *Futur. Oncol.* **2021**, *17*, 4405–4413. [CrossRef]

45. Huq, S.; Khalafallah, A.M.; Ruiz-Cardozo, M.A.; Botros, D.; Oliveira, L.A.; Dux, H.; White, T.; Jimenez, A.E.; Gujar, S.K.; Sair, H.I.; et al. A Novel Radiographic Marker of Sarcopenia with Prognostic Value in Glioblastoma. *Clin. Neurol. Neurosurg.* **2021**, *207*, 106782. [CrossRef] [PubMed]

46. Furtner, J.; Berghoff, A.S.; Schöpf, V.; Reumann, R.; Pascher, B.; Woitek, R.; Asenbaum, U.; Pelster, S.; Leitner, J.; Widhalm, G.; et al. Temporal muscle thickness is an independent prognostic marker in melanoma patients with newly diagnosed brain metastases. *J. Neuro-Oncol.* **2018**, *140*, 173–178. [CrossRef]

47. Ilic, I.; Faron, A.; Heimann, M.; Potthoff, A.-L.; Schäfer, N.; Bode, C.; Borger, V.; Eichhorn, L.; Giordano, F.; Güresir, E.; et al. Combined Assessment of Preoperative Frailty and Sarcopenia Allows the Prediction of Overall Survival in Patients with Lung Cancer (NSCLC) and Surgically Treated Brain Metastasis. *Cancers* **2021**, *13*, 3353. [CrossRef]

48. Leone, R.; Sferruzza, G.; Calimeri, T.; Steffanoni, S.; Conte, G.; De Cobelli, F.; Falini, A.; Ferreri, A.; Anzalone, N. Quantitative muscle mass biomarkers are independent prognosis factors in primary central nervous system lymphoma: The role of L3-skeletal muscle index and temporal muscle thickness. *Eur. J. Radiol.* **2021**, *143*, 109945. [CrossRef]

49. Mi, E.; Mauricaite, R.; Pakzad-Shahabi, L.; Chen, J.; Ho, A.; Williams, M. Deep learning-based quantification of temporalis muscle has prognostic value in patients with glioblastoma. *Br. J. Cancer* **2021**, *126*, 196–203. [CrossRef] [PubMed]

Article

Weight Change during the Early Phase of Convalescent Rehabilitation after Stroke as a Predictor of Functional Recovery: A Retrospective Cohort Study

Hiroshi Kishimoto [1,*], Yuka Nemoto [2], Takayuki Maezawa [3], Kazushi Takahashi [3], Kazunori Koseki [3], Kiyoshige Ishibashi [3], Hanako Tanamachi [3], Naoki Kobayashi [3] and Yutaka Kohno [4]

[1] Department of Physical Medicine and Rehabilitation, Ibaraki Prefectural University of Health Sciences Hospital, Ibaraki 300-0331, Japan
[2] Department of Nutritional Management, Ibaraki Prefectural University of Health Sciences Hospital, Ibaraki 300-0331, Japan; negisan0523@gmail.com
[3] Department of Physical Therapy, Ibaraki Prefectural University of Health Sciences Hospital, Ibaraki 300-0331, Japan; maesawat@ami.ipu.ac.jp (T.M.); takahashik@ami.ipu.ac.jp (K.T.); koseki@ami.ipu.ac.jp (K.K.); ishibashik@ami.ipu.ac.jp (K.I.); takanoh@ami.ipu.ac.jp (H.T.); kobayashin@ami.ipu.ac.jp (N.K.)
[4] Department of Neurology, Ibaraki Prefectural University of Health Sciences Hospital, Ibaraki 300-0331, Japan; kohno@ipu.ac.jp
* Correspondence: kishimotoh@ipu.ac.jp

Abstract: It has been reported that weight gain at discharge compared with admission is associated with improved activities of daily living in convalescent rehabilitation (CR) patients with low body mass index. Here, we investigated whether weight maintenance or gain during the early phase of CR after stroke correlates with a better functional recovery in patients with a wide range of BMI values. We conducted this retrospective cohort study in a CR ward of our hospital and included adult stroke patients admitted to the ward from January 2014 to December 2018. After ~1 month of hospitalization, the patients were classified into weight loss and weight maintenance or gain (WMG) groups based on the Global Leadership Initiative on Malnutrition criteria for weight. We adopted the motor functional independence measure (FIM) gain as the primary outcome. The motor FIM gain tended to be greater in the WMG group but without statistical significance. However, multiple regression analysis showed that WMG was significantly and positively associated with motor FIM gain. In conclusion, weight maintenance or gain in patients during the early phase of CR after stroke may be considered as a predictor of their functional recovery, and nutritional management to prevent weight loss immediately after the start of rehabilitation would contribute to this.

Keywords: convalescent rehabilitation; stroke; body weight; functional recovery; nutritional management

Citation: Kishimoto, H.; Nemoto, Y.; Maezawa, T.; Takahashi, K.; Koseki, K.; Ishibashi, K.; Tanamachi, H.; Kobayashi, N.; Kohno, Y. Weight Change during the Early Phase of Convalescent Rehabilitation after Stroke as a Predictor of Functional Recovery: A Retrospective Cohort Study. *Nutrients* **2022**, *14*, 264. https://doi.org/10.3390/nu14020264

Academic Editor: Yoshihiro Yoshimura

Received: 29 November 2021
Accepted: 5 January 2022
Published: 9 January 2022

Publisher's Note: MDPI stays neutral with regard to jurisdictional claims in published maps and institutional affiliations.

1. Introduction

The aging of the society is a global challenge today, and frailty [1] has become a critical issue. While there have been many molecular biological [2] and biochemical [3,4] studies on frailty, various epidemiological studies have also been conducted, and it is known that people with frailty have a higher risk of developing strokes [5], as well as a higher risk of falls and hip fractures [6]. In Japan, the national health insurance system established a convalescent rehabilitation (CR) ward in the year 2000, which has played an important role in the post-acute care of patients with stroke, brain or spinal cord injury, hip fracture, and hospital-associated deconditioning [7]. It has been reported that the prevalence of malnutrition, malnutrition risk status, and sarcopenia is high among patients admitted to the ward [8,9]. In addition, having malnutrition or sarcopenia has been associated with poor recovery of physical function in CR [9,10]. On the other hand, improvement of nutritional status among malnourished elderly patients with stroke during CR has been linked to improved activities of daily living (ADLs) [11,12]. Furthermore, we have previously reported that a group of patients whose nutritional status was maintained at good or even slightly improved from poor during CR had better functional recovery than a group of patients whose nutritional status remained poor or worsened even if their status was good at admission [13].

Among these studies, MNA -SF [14] in Ref. 11 GNRI [15] in Ref. 12, and CONUT [16] in Ref. 13 have been used as diagnostic tools for malnutrition or monitoring indicators of nutritional status, and recently, "GLIM(Global Leadership Initiative on Malnutrition) criteria for the diagnosis of malnutrition" [17] has been proposed as a consensus report from several clinical nutrition societies around the world. The GLIM criteria state that screening for nutritional status should be conducted using validated tools such as the MNA-SF [14], NRS-2002 [18], MUST [19], and SGA [20] and that priority should be given to repeated weight measurements over time to identify trajectories of weight loss, maintenance, and improvement. The importance of recognizing the pace of weight loss in the early stages of illness or injury has been emphasized in GLIM criteria [17]. One study that focused on weight change in CR was conducted by Kokura et al. [21]. They reported that in CR patients with a low body mass index (BMI) at admission, weight gain over the entire hospital stay up to the time of discharge was associated with improved ADL [21].

As in the case of frailty, molecular biological analysis has been conducted on stroke patients [22], but many epidemiological studies have also been conducted [23–25]. The range of overweight not reaching obesity is also considered to be a risk factor for stroke [24], and it has been shown that there are not a few stroke rehabilitation patients with high BMI [25]. However, the relationship between weight change in CR and improvement of ADL at the time of CR discharge in patients with a wide range of BMI has not been clarified so far. In addition, we have not found any studies that have assessed weight change in the early stages of CR and examined its relationship with functional recovery.

This study aims to address the clinical question of whether weight maintenance gain or loss in the early stages of CR in patients with stroke has a positive or negative impact, respectively, on functional recovery at discharge.

2. Materials and Methods

2.1. Participants and Setting

We conducted a retrospective cohort study in a 47-bed CR ward of our hospital in Japan. Adult stroke patients admitted to the ward from January 2014 to December 2018 were included in the study. In this ward, the daily program consisted of physical therapy (PT), occupational therapy (OT), and speech and hearing therapy (ST), with a maximum of 180 min per day (according to the Japanese health insurance system). On average, about 150 min were provided per day, including 40–120 min of PT, 40–120 min of OT, and 0–60 min of ST. The training content is the same as in previous literatures [26,27]. Patients with a hospital stay of fewer than 30 days after cohorting were excluded due to a short observation period. We excluded patients with a BMI of 30 or higher on admission because they are treated as obese under the Japanese health insurance system and are offered a diet aimed at weight loss. Patients with missing data were also excluded.

2.2. Cohorting

Based on the GLIM criteria for weight loss, "unintentional weight loss of >5% within the past 6 months" [17], the weight at admission was subtracted from the weight on day N, ~1 month after admission, and then multiplied by 180/N to convert to a change per 180 days. Patients whose converted decrease was >5% of their original weight were classified as the weight loss (WL) group and the remaining patients formed the weight maintenance or gain (WMG) group.

2.3. Data Collection

Baseline patients' characteristics and parameters such as age, gender, BMI, stroke type, number of days between the onset and admission to our rehabilitation ward, serum albumin level, serum creatinine level, and motor/cognitive functional independence measure (FIM) scores at admission [28] were obtained from the medical records retrospectively. The presence of dysphagia was defined by tube feeding or provision of the texture-modified meal to support the swallowing function. The severity of complications was assessed using the Charlson comorbidity index (CCI) [29]. Energy intake (EI; kcal) was determined from the ratio of the patient's food intake to the food supply recorded visually by the nurse or dietitian and averaged over the three days prior to cohorting. The ratio of EI and basal energy expenditure (BEE), calculated from the Harris–Benedict equation [30], was also recorded for each patient. The number of days spent in the hospital and the rehabilitation minutes per day during the hospital stay were also recorded.

2.4. Outcome

We adopted the motor FIM [28] gain as the primary outcome of the study. The FIM consists of 13 motor items and five cognitive items. Each item is rated on a scale of 1–7, with motor items scored from 13 to 91 and cognitive items scored from 5 to 35. The motor FIM gain was calculated as the difference in motor FIM score between the discharge and admission.

2.5. Sample Size Calculation

Statistical software G*power 3 [31] was used for sample size calculation.

Multiple regression analysis using FIM efficiency as the outcome in our previous study [13] yielded an effect size of f2 = 0.159. Using the effect size of 0.159, significance level of $\alpha = 0.05$, power of test $1 - \beta = 0.9$, assuming the number of planned explanatory variables to be 13, the required sample size was calculated to be 153. Since the number of patients admitted to our CR ward has ranged from 60 to 90 per year, we decided to obtain statistics for 5 years by considering the exclusion criteria and missing data.

2.6. Statistical Analysis

Mann–Whitney U test, t test, chi-square test, and Fisher's exact probability test were used to compare the two groups, depending on the type of variable. The Shapiro–Wilk test was performed to assess normality. To examine the association between WMG/WL and the outcome, we applied multiple regression analysis. To avoid the effects of confounding variables, additional explanatory variables included age, sex, type of stroke, number of days since onset, BMI, serum albumin level, serum creatinine level, motor FIM at admission, CCI, presence of dysphagia, duration of hospital stay, and rehabilitation time per day. Since EI/BEE was expected to be strongly related to WMG, we performed an additional multiple regression analysis (Model 2) using EI/BEE instead of WMG. Multicollinearity was assessed by the variance inflation factor (VIF), and if the VIF value was less than 10, it was considered that no multicollinearity was observed. The significance level was set at a p-value < 0.05 and all the analyses were performed using statistical software SPSS Statistics, version 26 (IBM Japan, Tokyo Japan).

3. Results

Of the 393 patients admitted to the CR ward during the 5-year period, 100 were excluded from the study, resulting in 293 patients for analysis, divided into the WMG group ($n = 176$) and the WL group ($n = 117$) (Figure 1). Table 1 shows the patient characteristics. There were significant differences between the two groups in age, number of days from onset to admission to the convalescent rehabilitation ward, BMI, serum albumin level at admission, EI, and EI/BEE. The motor FIM gain as the outcome of the study was greater in the WMG group, but it did not reach statistical significance.

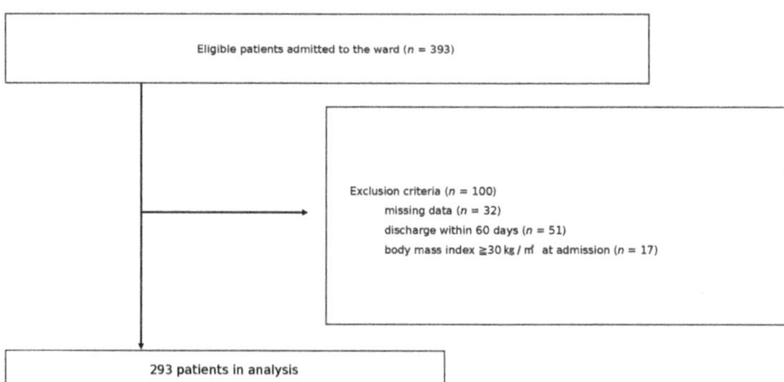

Figure 1. Study flowchart.

Table 1. Patients' characteristics.

	Total (*N* = 293)	WMG (*N* = 176)	WL (*N* = 117)	*p* Value	
Age, years, median [IQR]	69 [60–78]	70 [61–79]	67 [56–76.5]	0.041	*
Sex: Male, *n* (%)	178 (60.8)	99 (55.6)	79 (44.4)		
Female, *n* (%)	115 (39.2)	77 (67.0)	38 (33.0)	0.067	†
Stroke type					
Cerebral infarction, *n* (%)	151 (51.5)	85 (56.3)	66 (43.7)		
Intracerebral hemorrhage, *n* (%)	119 (40.6)	75 (63.0)	44 (37.0)	0.333	‡
Subarachnoid hemorrhage, *n* (%)	23 (7.8)	16 (69.6)	7 (30.4)		
Days from onset to admission, d, median [IQR]	36 [27.5–50]	38 [30–51.75]	35 [24–42.5]	0.007	*
Days from admission to cohorting, d, median [IQR]	28 [21–35]	28 [21.75–35]	28 [21–34]	0.482	*
Serum albumin level on admission, mg/dL, median [IQR]	3.9 [3.5–4.2]	3.85 [3.4–4.2]	4.1 [3.7–4.3]	0.001	*
Serum creatinine level on admission, mg/dL, median [IQR]	0.7 [0.6–0.9]	0.7 [0.6–0.9]	0.7 [0.6–0.9]	0.321	*
CCI, median, [IQR]	2 [2–3]	2 [2–3]	2 [2–3]	0.969	*
BMI at admission, kg/m², mean (SD)	22.1 ± 3.1	21.4 ± 3.0	23.0 ± 2.9	<0.001	§
Low BMI at admission (GLIM criteria for Asians)					
Yes, *n*, (%)	50 (17.1)	40 (22.7)	10 (8.5)		
No, *n*, (%)	243 (82.9)	136 (77.3)	107 (91.5)	0.001	†
BMI at discharge, kg/m², mean (SD)	21.7 ± 2.7	21.6 ± 2.8	21.8 ± 2.6	0.587	§
Energy intake, kcal/kg BW/day, mean (SD)	26.5 ± 5.8	28.0 ± 5.8	24.2 ± 5.8	<0.001	§
Protein intake, g/kg BW/day, median [IQR]	1.12 [0.97–1.28]	1.19 [1.05–1.36]	1.05 [0.92–1.17]	<0.001	*
Energy intake/Basal energy expenditure, median [IQR]	1.26 [1.11–1.38]	1.30 [1.18–1.41]	1.16 [1.00–1.30]	<0.001	*
Rehabilitation therapy, min/day, median [IQR]	137 [120–147]	137 [118–148]	137 [121–146]	0.402	*
FIM score on admission, median [IQR]					
Total FIM	73 [49–96.5]	70 [45–92]	78 [53–101]	0.088	*
Motor FIM	49 [28–68.5]	46 [26.25–65]	55 [35–71]	0.073	*
Cognitive FIM	24 [17–30.5]	24 [16–30]	24 [17–31]	0.342	*
Length of hospital stay, days, median [IQR]	124 [96–158]	130.5 [100–155.75]	120 [92.5–166.5]	0.626	*
FIM gain, median [IQR]	24 [15–37]	25.5 [15–38]	23 [15–36]	0.298	*
Motor FIM gain, median [IQR]	20 [11–30.5]	21 [12–31]	19 [10–27.5]	0.232	*

* Mann–Whitney U test; † Fisher's exact test; ‡ Chi-square test; § *t* test; WMG: weight maintenance or gain group; WL: weight loss group; IQR: interquartile range; SD: standard deviation; CCI: Charlson comorbidity index; BMI: body mass index; EI/BEE: energy intake (kcal)/Basal energy expenditure (kcal); FIM: functional independence measure.

The results of multiple regression analysis with motor FIM gain as the dependent variable and WMG as an explanatory variable showed that WMG was significantly and positively associated with motor FIM gain (standardized coefficient = 0.105, p = 0.043, adjusted R-square = 0.320). In addition, age, days from onset to admission to the convalescent rehabilitation ward, motor FIM at admission, CCI, and dysphagia showed significant negative associations, whereas BMI and hospital stay showed significant positive associations (Table 2).

Table 2. Multivariate analysis of motor FIM gain (model 1).

Factor	Standardized Coefficient	p-Value	VIF
Age	−0.125	0.037	1.526
Sex	0.031	0.590	1.449
Stroke type	0.009	0.878	1.305
Days from onset	−0.212	<0.001	1.449
BMI	0.133	0.011	1.142
Serum albumin level on admission, mg/dL	0.059	0.360	1.779
Serum creatinine level on admission, mg/dL	−0.027	0.638	1.417
Motor FIM at admission	−0.489	<0.001	2.317
CCI	−0.126	0.019	1.217
Dysphagia	−0.166	0.006	1.552
Length of stay	0.208	<0.001	1.404
Rehabilitation therapy, min/day	0.082	0.119	1.181
WMG	0.106	0.043	1.156

Multiple regression analysis (model 2, Table 3), which included motor FIM gain as the outcome and EI/BEE instead of WMG as an explanatory variable, showed that EI/BEE was significantly and positively associated with motor FIM gain (standardized coefficient = 0.169, p = 0.005, adjusted R-square = 0.329). In both models, no variable had a VIF > 10.

Table 3. Multivariate analysis of motor FIM gain (model 2).

Factor	Standardized Coefficient	p-Value	VIF
Age	−0.197	0.003	1.869
Sex	0.018	0.756	1.442
Stroke type	0.008	0.885	1.305
Days from onset	−0.208	0.000	1.431
BMI	0.170	0.002	1.284
Serum albumin level on admission, mg/dL	0.037	0.559	1.747
Serum creatinine level on admission, mg/dL	−0.021	0.715	1.414
motor FIM at admission	−0.521	<0.001	2.374
CCI	−0.121	0.024	1.220
Dysphagia	−0.142	0.021	1.614
Length of stay	0.194	<0.001	1.420
Rehabilitation therapy, min/day	0.072	0.168	1.190
EI/BEE	0.169	0.005	1.529

4. Discussion

In comparing the two groups, the WMG group had a greater motor FIM gain than the WL group, but the difference was not statistically significant. However, after adjusting for confounding factors, such as age, sex, type of stroke, number of days since onset, BMI, serum albumin level, serum creatinine level, motor FIM at admission, Charlson comorbidity index, presence of dysphagia, duration of hospital stay, and rehabilitation time per day, in a multiple regression analysis, WMG showed a significant positive association with motor FIM gain.

Instead of weight gain during the hospitalization period of convalescent patients in the previous study [21], our study showed that WMG during the early stage of convalescent rehabilitation after stroke was associated with better functional recovery. It is also noteworthy that the study included not only underweight patients but also patients with a wide range of body weight, excluding only obese patients who needed weight loss guidance.

Although it is unknown whether functional outcomes in adults undergoing inpatient stroke rehabilitation are affected by obesity [25], in this study, BMI was significantly positively associated with motor FIM gain in the group of patients with a BMI < 30.

Multiple regression analysis (model 2), where EI/BEE was used as an explanatory variable instead of WMG, revealed that EI/BEE was also significantly positively associated with motor FIM gain. These results are supported by Nii et al. who reported that nutritional intake in the early stage of admission to a rehabilitation ward is associated with FIM efficiency [12]. Shimazu et al. reported that frequent individualized nutritional support was associated with improved nutritional status, physical function, and dysphagia after stroke, which emphasizes the need for intensive multidisciplinary nutritional support [32]. These findings and the results of our study suggest that in stroke convalescent rehabilitation, frequent nutritional support to maintain or increase weight should be provided by the team in charge of each patient, ensuring adequate EI by constantly considering the increased energy expenditure to rehabilitation. In such support, the median EI/BEE of 1.30 in the WMG group could be considered as one of the guidelines for energy supply in stroke rehabilitation.

Besides being retrospective and single-center study, our study is also limited by the fact that motor FIM gain was used as the outcome, so functional recovery was only captured from aspects that can be assessed by motor FIM, and that muscle mass change may be more strongly associated with motor FIM gain compared with weight change. Although body weight measurements are more readily available compared with muscle mass measurements, a recent study [33] has shown that muscle mass gain could be positively associated with functional recovery in patients with sarcopenia after stroke. Future studies should focus on muscle mass change as well.

5. Conclusions

In conclusion, weight maintenance or gain in patients during the early phase of hospitalized convalescent rehabilitation after stroke was a predictor of functional recovery in patients with a BMI of <30. Therefore, nutritional management to prevent weight loss immediately after the start of convalescent rehabilitation is advised to ensure better functional recovery in these patients.

Author Contributions: Conceptualization, H.K., Y.N., K.T., K.K., K.I., H.T. and N.K.; methodology, H.K. and T.M.; software, H.K. and T.M.; validation, K.T., K.K., K.I., H.T. and N.K.; formal analysis, H.K.; investigation, H.K. and Y.N.; data curation, H.K., T.M., Y.N., K.T., K.K., K.I., H.T. and N.K.; writing—original draft preparation, H.K. and Y.N.; writing—review and editing, Y.N., K.T., K.K., K.I., H.T., N.K. and Y.K.; visualization, H.K. and H.T.; supervision, Y.K.; project administration, H.K. and Y.N. All authors have read and agreed to the published version of the manuscript.

Funding: This study received an FY2020 incentive research grant from Ibaraki Prefectural University of Health Sciences.

Institutional Review Board Statement: The study was conducted according to the guidelines of the Declaration of Helsinki and was approved by the Ethics Committee of Ibaraki Prefectural University of Health Sciences (protocol code e276, 25 August 2020).

Informed Consent Statement: Because this was a retrospective study, written informed consent for the study was not obtained from each participant. However, after the start of our study, the opt-out was publicly posted in the hospital and on the Internet, and procedures were in place to allow participants to withdraw from the study at any time.

Data Availability Statement: No new data were created or analyzed in this study. Data sharing is not applicable to this article.

Conflicts of Interest: The authors declare no conflict of interest.

References

1. Fried, L.P.; Tangen, C.M.; Walston, J.; Newman, A.B.; Hirsch, C.; Gottdiener, J.; Seeman, T.; Tracy, R.; Kop, W.J.; Burke, G.; et al. Frailty in older adults: Evidence for a phenotype. *J. Gerontol. A Biol. Sci. Med. Sci.* **2001**, *56*, M146–M156. [CrossRef]
2. Lorenzi, M.; Bonassi, S.; Lorenzi, T.; Giovannini, S.; Bernabei, R.; Onder, G. A review of telomere length in sarcopenia and frailty. *Biogerontology* **2018**, *19*, 209–221. [CrossRef]
3. Giovannini, S.; Onder, G.; Lattanzio, F.; Bustacchini, S.; Stefano, G.D.; Moresi, S.; Russo, A.; Bernabei, R.; Landi, F. Selenium Concentrations and Mortality Among Community-Dwelling Older Adults: Results from ilSIRENTE Study. *J. Nutr. Health Aging* **2018**, *22*, 608–612. [CrossRef]
4. Giovannini, S.; Onder, G.; Leeuwenburgh, C.; Carter, C.; Marzetti, E.; Russo, A.; Capoluongo, E.; Pahor, M.; Bernabei, R.; Landi, F. Myeloperoxidase levels and mortality in frail community-living elderly individuals. *J. Gerontol. A Biol. Sci. Med. Sci.* **2010**, *65*, 369–376. [CrossRef] [PubMed]
5. Gugganig, R.; Aeschbacher, S.; Leong, D.P.; Meyre, P.; Blum, S.; Coslovsky, M.; Beer, J.H.; Moschovitis, G.; Müller, D.; Anker, D.; et al. Frailty to predict unplanned hospitalization, stroke, bleeding, and death in atrial fibrillation. *Eur. Heart J. Qual. Care Clin. Outcomes* **2021**, *25*, 42–51. [CrossRef] [PubMed]
6. Ensrud, K.E.; Ewing, S.K.; Taylor, B.C.; Fink, H.A.; Cawthon, P.M.; Stone, K.L.; Hillier, T.A.; Cauley, J.A.; Hochberg, M.C.; Rodondi, N.; et al. Comparison of 2 frailty indexes for prediction of falls, disability, fractures, and death in older women. *Arch. Intern. Med.* **2008**, *168*, 382–389. [CrossRef] [PubMed]
7. Miyai, I.; Sonoda, S.; Nagai, S.; Takayama, Y.; Inoue, Y.; Kakehi, A.; Kurihara, M.; Ishikawa, M. Results of new policies for inpatient rehabilitation coverage in Japan. *Neurorehabil. Neural Rep.* **2011**, *25*, 540–547. [CrossRef] [PubMed]
8. Wakabayashi, H.; Sakuma, K. Rehabilitation nutrition for sarcopenia with disability: A combination of both rehabilitation and nutrition care management. *J. Cachexia Sarcopenia Muscle.* **2014**, *5*, 269–277. [CrossRef]
9. Nishioka, S.; Wakabayashi, H.; Kayashita, J.; Taketani, Y.; Momosaki, R. Predictive validity of the Mini Nutritional Assessment Short-Form for rehabilitation patients: A retrospective analysis of the Japan Rehabilitation Nutrition Database. *J. Hum. Nutr. Diet.* **2021**, *34*, 881–889. [CrossRef]
10. Yoshimura, Y.; Wakabayashi, H.; Bise, T.; Nagano, F.; Shimazu, S.; Shiraishi, A.; Yamaga, M.; Koga, H. Sarcopenia is associated with worse recovery of physical function and dysphagia and a lower rate of home discharge in Japanese hospitalized adults undergoing convalescent rehabilitation. *Nutrition* **2019**, *61*, 111–118. [CrossRef]
11. Nishioka, S.; Wakabayashi, H.; Nishioka, E.; Yoshida, T.; Mori, N.; Watanabe, R. Nutritional improvement correlates with recovery of activities of daily living among malnourished elderly stroke patients in the convalescent stage: A cross-sectional study. *J. Acad. Nutr. Diet.* **2016**, *116*, 837–843. [CrossRef]
12. Nii, M.; Maeda, K.; Wakabayashi, H.; Tanaka, A. Nutritional improvement and energy intake are associated with functional recovery in patients after cerebrovascular disorders. *J. Stroke Cerebrovasc. Dis.* **2016**, *25*, 57–62. [CrossRef]
13. Kishimoto, H.; Yozu, A.; Kohno, Y.; Oose, H. Nutritional improvement is associated with better functional outcome in stroke rehabilitation: A cross-sectional study using controlling nutritional status. *J. Rehabil. Med.* **2020**, *52*, jrm00029. [CrossRef]
14. Rubenstein, L.Z.; Harker, J.O.; Salvà, A.; Guigoz, Y.; Vellas, B. Screening for undernutrition in geriatric practice: Developing the short- form mini-nutritional assessment (MNA- SF). *J. Gerontol. A Biol. Sci. Med. Sci.* **2001**, *56*, M366–M372. [CrossRef] [PubMed]
15. Cereda, E.; Vanotti, A. The new Geriatric Nutritional Risk Index is a good predictor of muscle dysfunction in institutionalized older patients. *Clin. Nutr.* **2007**, *26*, 78–83. [CrossRef] [PubMed]
16. Ignacio de Ulíbarri, J.; González-Madroño, A.; de Villar, N.G. CONUT: A tool for controlling nutritional status. First validation in a hospital population. *Nutr. Hosp.* **2005**, *20*, 38–45.
17. Cederholm, T.; Jensen, G.L.; Correia, M.I.T.D.; Gonzalez, M.C.; Fukushima, R.; Higashiguchi, T.; Baptista, G.; Barazzoni, R.; Blaauw, R.; Coats, A.J.; et al. GLIM criteria for the diagnosis of malnutrition - A consensus report from the global clinical nutrition community. *Clin. Nutr.* **2019**, *38*, 1–9. [CrossRef] [PubMed]
18. Kondrup, J.; Allison, S.P.; Elia, M.; Vellas, B.; Plauth, M. Educational and Clinical Practice Committee, European Society of Parenteral and Enteral Nutrition (ESPEN). ESPEN guidelines for nutrition screening 2002. *Clin. Nutr.* **2003**, *22*, 415–421. [CrossRef]
19. Stratton, R.J.; Hackston, A.; Longmore, D.; Dixon, R.; Price, S.; Stroud, M.; King, C.; Elia, M. Malnutrition in hospital outpatients and inpatients: Prevalence, concurrent validity and ease of use of the 'malnutrition universal screening tool' ('MUST') for adults. *Br. J. Nutr.* **2004**, *92*, 799–808. [CrossRef]
20. Detsky, A.S.; McLaughlin, J.R.; Baker, J.P.; Johnston, N.; Whittaker, S.; Mendelson, R.A.; Jeejeebhoy, K.N. What is subjective global assessment of nutritional status? *J. Parenter. Enter. Nutr.* **1987**, *11*, 8–13. [CrossRef]
21. Kokura, Y.; Nishioka, S.; Okamoto, T.; Takayama, M.; Miyai, I. Weight gain is associated with improvement in activities of daily living in underweight rehabilitation inpatients: A nationwide survey. *Eur. J. Clin. Nutr.* **2019**, *73*, 1601–1604. [CrossRef]
22. Biscetti, F.; Giovannini, S.; Straface, G.; Bertucci, F.; Angelini, F.; Porreca, C.; Landolfi, R.; Flex, A. RANK/RANKL/OPG pathway: Genetic association with history of ischemic stroke in Italian population. *Eur. Rev. Med. Pharmacol. Sci.* **2016**, *20*, 4574–4580.
23. Krishnamurthi, R.V.; Feigin, V.L.; Forouzanfar, M.H.; Mensah, G.; Connor, M.; Bennett, D.; Moran, A.; Sacco, R.L.; Anderson, L.M.; Truelsen, T.; et al. Global and regional burden of first-ever ischaemic and haemorrhagic stroke during 1990–2010: Findings from the Global Burden of Disease Study 2010. *Lancet Glob. Health* **2013**, *1*, e259–e281. [CrossRef]
24. Strazzullo, P.; D'Elia, L.; Cairella, G.; Garbagnati, F.; Cappuccio, F.P.; Scalfi, L. Excess body weight and incidence of stroke: Meta-analysis of prospective studies with 2 million participants. *Stroke* **2010**, *41*, e418–e426. [CrossRef]

25. Macdonald, S.; Journeay, W.; Uleryk, E. A systematic review of the impact of obesity on stroke inpatient rehabilitation functional outcomes. *NeuroRehabilitation* **2020**, *46*, 403–415. [CrossRef] [PubMed]

26. Umehara, T.; Tanaka, R.; Tsunematsu, M.; Sugihara, K.; Moriuchi, Y.; Yata, K.; Muranaka, K.; Inoue, J.; Kohriyama, T.; Kakehashi, M. Can the Amount of Interventions during the Convalescent Phase Predict the Achievement of Independence in Activities of Daily Living in Patients with Stroke? A Retrospective Cohort Study. *J. Stroke Cerebrovasc. Dis.* **2018**, *27*, 2436–2444. [CrossRef]

27. Yoshimura, Y.; Wakabayashi, H.; Nagano, F.; Bise, T.; Shimazu, S.; Shiraishi, A. Low Hemoglobin Levels are Associated with Sarcopenia, Dysphagia, and Adverse Rehabilitation Outcomes After Stroke. *J. Stroke Cerebrovasc. Dis.* **2020**, *29*, 105405. [CrossRef]

28. Ottenbacher, K.J.; Hsu, Y.; Granger, C.V.; Fiedler, R.C. The reliability of the functional independence measure: A quantitative review. *Arch. Phys. Med. Rehabil.* **1996**, *77*, 1226–1232. [CrossRef]

29. Charlson, M.E.; Pompei, P.; Ales, K.L.; MacKenzie, C.R. A new method of classifying prognostic comorbidity in longitudinal studies: Development and validation. *J. Chronic. Dis.* **1987**, *40*, 373–383. [CrossRef]

30. Harris, J.A.; Benedict, F.G. *A Biometric Study of Basal Metabolism in Man*; Carnegie Institute of Washington: Washington, DC, USA, 1919.

31. Faul, F.; Erdfelder, E.; Buchner, A.; Lang, A.G. G*Power 3: Statistical power analyses using G*Power 3.1: Tests for correlation and regression analyses. *Behav. Res. Methods* **2009**, *41*, 1149–1160. [CrossRef]

32. Shimazu, S.; Yoshimura, Y.; Kudo, M.; Nagano, F.; Bise, T.; Shiraishi, A. Frequent and personalized nutritional support leads to improved nutritional status, activities of daily living, and dysphagia after stroke. *Nutrition* **2021**, *83*, 111091. [CrossRef] [PubMed]

33. Nagano, F.; Yoshimura, Y.; Bise, T.; Shimazu, S.; Shiraishi, A. Muscle mass gain is positively associated with functional recovery in patients with sarcopenia after stroke. *J. Stroke Cerebrovasc. Dis.* **2020**, *29*, 105017. [CrossRef] [PubMed]

 *nutrients*

Article

Relationships between Pre-Stroke SARC-F Scores, Disability, and Risk of Malnutrition and Functional Outcomes after Stroke—A Prospective Cohort Study

Masafumi Nozoe [1],*, Hiroki Kubo [2], Masashi Kanai [1] and Miho Yamamoto [2]

[1] Department of Physical Therapy, Faculty of Nursing and Rehabilitation, Konan Women's University, Kobe 658-0001, Japan; kanaimasa07@gmail.com
[2] Department of Rehabilitation, Itami Kousei Neurosurgical Hospital, Itami 664-0028, Japan; hiro.k16862@gmail.com (H.K.); bf39my23@gmail.com (M.Y.)
* Correspondence: masafumi.nozoe@gmail.com; Tel.: +81-78-413-3584

Abstract: SARC-F is a screening tool for sarcopenia; however, it has not yet been established whether SARC-F scores predict functional outcomes. Therefore, we herein investigated the relationship between SARC-F scores and functional outcomes in stroke patients. The primary outcome in the present study was the modified Rankin Scale (mRS) 3 months after stroke. The relationship between SARC-F scores and poor functional outcomes was examined using a logistic regression analysis. Furthermore, the applicability of SARC-F scores to the assessment of poor functional outcomes was analyzed based on the area under the receiver operating curve (ROC). Eighty-one out of the 324 patients enrolled in the present study (25%) had poor functional outcomes (mRS $\geq$ 4). The results of the multivariate analysis revealed a correlation between SARC-F scores (OR = 1.29, 95% CI = 1.05–1.59, $p = 0.02$) and poor functional outcomes. A cut-off SARC-F score $\geq$ 4 had low-to-moderate sensitivity (47.4%) and high specificity (87.3%). The present results suggest that the measurement of pre-stroke SARC-F scores is useful for predicting the outcomes of stroke patients.

Keywords: stroke; sarcopenia; SARC-F score; disability; malnutrition risks

Citation: Nozoe, M.; Kubo, H.; Kanai, M.; Yamamoto, M. Relationships between Pre-Stroke SARC-F Scores, Disability, and Risk of Malnutrition and Functional Outcomes after Stroke—A Prospective Cohort Study. *Nutrients* **2021**, *13*, 3586. https://doi.org/10.3390/nu13103586

Academic Editor: Yoshihiro Yoshimura

Received: 27 August 2021
Accepted: 9 October 2021
Published: 13 October 2021

Publisher's Note: MDPI stays neutral with regard to jurisdictional claims in published maps and institutional affiliations.

1. Introduction

Stroke is one of the leading causes of disability [1], and the primary cause of disability in stroke patients is brain injury [2]. Stroke increases the risk of sarcopenia [3], which is a contributing factor to disability in these patients [4]. The prevalence of stroke-related sarcopenia increases in the elderly and is associated with the poor recovery of activities of daily living [5,6]. Although sarcopenia may develop in stroke patients, it may also be present pre-stroke, and has been implicated in poor functional outcomes [7–9].

SARC-F is a screening tool for sarcopenia in elderly subjects [9–12] and predicts the risk of pre-stroke sarcopenia [9]. A previous study reported that the risk of developing sarcopenia was elevated in patients with a SARC-F score $\geq$4 [13]. Since SARC-F has low sensitivity, but high specificity [11,14], the risk of sarcopenia assessed by SARC-F may be underestimated [11,14–16]. Another cut-off score was recently reported for predicting the risk of sarcopenia (SARC-F = 1) [17] and the usefulness of SARC-F for detecting the risk of frailty or falls has been demonstrated [18,19]. A previous study revealed a relationship between SARC-F scores and muscle mass [20,21]. Based on these findings, SARC-F scores may be useful for predicting the risk of pre-stroke sarcopenia and functional outcomes in stroke patients.

The aim of the present study was to investigate the relationship between pre-stroke SARC-F scores and functional outcomes in stroke patients. We hypothesized that SARC-F scores are associated with functional outcomes, even after adjustments for confounding factors, such as pre-stroke disability or a risk of malnutrition.

2. Materials and Methods

2.1. Study Design and Participants

This prospective cohort study was conducted between August 2017 and December 2019 and included elderly patients consecutively admitted to Itami Kousei Neurosurgical Hospital within 48 h of stroke onset. The following patients were enrolled: age $\geq$ 65 years and evidence of cerebral infarction or intracerebral hemorrhage on computed tomography or magnetic resonance imaging. Exclusion criteria were (1) pre-stroke dependent ambulation, (2) patients unable to complete the questionnaire because of impaired consciousness, cognitive dysfunction, or language disorders, such as aphasia, and (3) the lack of informed consent. The present study was approved by the Research Ethics Committee of Konan Women's University, and all patients provided their informed consent.

2.2. SARC-F

The SARC-F questionnaire, which evaluates the pre-stroke status and has been adapted for a Japanese population, was completed by patients within 5 days of admission. It comprises the following components: strength, assistance with walking, rising from a chair, climbing stairs, and falls [12]. SARC-F scores range between 0 and 10 (0 = best, 10 = worst), with each component receiving 0–2 points.

2.3. Assessment of a Risk of Malnutrition and Comorbidities

The Geriatric Nutritional Risk Index (GNRI) was employed to evaluate the nutritional status of patients [22], and was calculated as follows: GNRI = (1.489 $\times$ serum albumin [g/dL]) + 103 (41.7 $\times$ weight [kg]/ideal body weight). In cases in which weight/ideal body weight was $\geq$1.0, the ratio was set to 1. GNRI $\leq$ 98 kg/m^2 was defined as a risk of malnutrition as previously reported [22].

2.4. Clinical Characteristics

Information was obtained from electronic medical records on the following patient characteristics: age, sex, height, body weight, body mass index, neurological deficits evaluated by the National Institutes of Health Stroke Scale (NIHSS) score, stroke type, lesion laterality, and pre-stroke mRS. Pre-stroke disability was defined as mRS = 2 (slight disability) or 3 (moderate disability), and no disability as mRS = 0 (no symptoms) or 1 (no significant disability).

2.5. Main Outcome

The primary outcome of the present study was the modified Rankin Scale (mRS) evaluated 3 months after stroke from medical records or in a telephone interview. mRS was scored based on an unstructured direct interview by a physician [23], with 0 = no symptoms, 1 = no significant disability despite the presence of symptoms; capable of performing all of the usual duties and activities, 2 = slight disability; unable to perform all of the previous activities, but capable of attending to one's own affairs without assistance, 3 = moderate disability; requiring some help, but capable of walking without assistance, 4 = moderately severe disability; unable to walk or attend to one's own bodily needs without assistance, 5 = severe disability; bedridden, incontinent, and requiring constant nursing care and attention, and 6 = dead [24]. A poor outcome was defined as mRS scores 3 months after stroke of 4–6 [25].

2.6. Statistical Analysis

Data are shown as medians (interquartile range; IQR) and numbers (%) for categorical data. The unadjusted and adjusted odds ratios (OR) of SARC-F scores and poor functional outcomes (mRS $\geq$ 4) were calculated by a logistic regression analysis. Confounding factors were adjusted for, including age [26], sex [27], NIHSS [28], pre-stroke disability [29], and malnutrition risk [22,30]. The applicability of SARC-F scores to evaluations of poor functional outcomes was examined in an analysis of the area under the receiver operating curve (ROC). Sensitivity and 1-specificity were calculated from the obtained sensitivity and specificity, and the point at the maximal value was taken as the optimum cut-off value. The area under the curve, sensitivity, specificity, and positive and negative predictive values for SARC-F were analyzed. All statistical analyses were performed using SPSS version 20.0 (SPSS, Inc., Chicago, IL, USA). Differences with $p < 0.05$ were considered to be significant.

3. Results

During the study period, 643 elderly stroke patients were hospitalized, and 293 were excluded from the analysis due to admission 48 h after stroke symptom onset ($n = 17$), pre-stroke dependent ambulation ($n = 74$), impaired consciousness ($n = 68$), cognitive dysfunction ($n = 64$), and aphasia ($n = 39$). Ten patients also refused to participate and 21 did not provide informed consent. Among the 350 patients enrolled, 26 patients refused a follow-up. Therefore, 324 stroke patients were included in the present study.

Among the patients enrolled, 54 (17%) had pre-stroke disability and 41 (13%) were at risk of malnutrition. Eighty-one patients (25%) had poor functional outcomes (mRS $\geq$ 4) 3 months after stroke (Table 1). Table 2 shows the distribution for each SARC-F score in all patients. Approximately 50% of patients had an SARC-F score of zero.

Based on unadjusted OR, age (OR = 1.06, 95% confidence interval (CI) = 1.02–1.10, $p = 0.007$), NIHSS (OR = 1.40, 95% CI = 1.36–1.68, $p < 0.001$), pre-stroke disability (OR = 5.75, 95% CI = 3.00–10.99, $p < 0.001$), risk of malnutrition (OR = 3.78, 95% CI = 1.86–7.68, $p < 0.001$), and SARC-F scores (OR = 1.44, 95% CI = 1.27–1.64, $p < 0.001$) correlated with poor functional outcomes 3 months after stroke.

Table 1. Stroke patient characteristics.

	Total Cohort
	($n = 324$)
Age (years, median (IQR))	76 (11)
Sex (male/female)	187/137
Body mass index (kg/m^2, median [IQR])	22.5 (4.2)
NIHSS (median [IQR])	2 (3)
Stroke type (infarction/hemorrhage)	267/57
Lesion side (right/left/both)	161/152/11
Pre-stroke disability (%)	54 (17)
Stroke risk factors (%)	
Hypertension	159 (49)
Diabetes	74 (23)
Previous stroke	89 (28)
Hypercholesterolemia	78 (24)
Ischemic heart disease	27 (8)
Atrial fibrillation	27 (8)
Smoking	110 (34)
Risk of malnutrition (%)	41 (13)
mRS 3 months after stroke (%)	
0	31 (10)
1	105 (32)
2	58 (18)
3	49 (15)
4	69 (21)
5	10 (3)
6	2 (1)
Poor functional outcome (%)	81 (25)

IQR = interquartile range; NIHSS = National Institutes of Health Stroke Scale; mRS = modified Rankin Scale.

Table 2. SARC-F scores and their distribution.

	Total Cohort
	(*n* = 324)
SARC-F score (median [IQR])	1 (3)
SARC-F score distribution	
score = 0 (%)	158 (49)
score = 1 (%)	44 (13)
score = 2 (%)	31 (10)
score = 3 (%)	30 (9)
score = 4 (%)	30 (9)
score = 5 (%)	14 (4)
score = 6 (%)	2 (1)
score = 7 (%)	5 (2)
score = 8 (%)	7 (2)
score = 9 (%)	3 (1)
score = 10 (%)	0 (0)

IQR = interquartile range.

After adjustments for these factors, NIHSS (OR = 1.56, 95% CI = 1.39–1.76, $p < 0.001$), pre-stroke disability (OR = 3.22, 95% CI = 1.11–9.34, $p = 0.03$), and SARC-F scores (OR = 1.29, 95% CI = 1.05–1.59, $p = 0.02$) correlated with poor functional outcomes (Table 3).

Table 3. Logistic regression analysis of poor functional outcomes.

	Unadjusted		Adjusted	
	Odds Ratio (95% CI)	*p*-Value	Odds Ratio (95% CI)	*p*-Value
Age	1.06 (1.02–1.10)	0.007	1.03 (0.97–1.09)	0.41
Sex	1.4 (0.79–2.48)	0.25	0.73 (0.31–1.69)	0.46
NIHSS	1.51 (1.36–1.68)	<0.001	1.56 (1.39–1.76)	<0.001
Pre-stroke disability	5.75 (3.00–10.99)	<0.001	3.22 (1.11–9.34)	0.03
Risk of malnutrition	3.78 (1.86–7.68)	<0.001	2.5 (0.89–7.06)	0.08
SARC-F score	1.44 (1.27–1.64)	<0.001	1.29 (1.05–1.59)	0.02

CI = confidence interval; NIHSS = National Institutes of Health Stroke Scale.

The results of the ROC curve analysis are shown in Figure 1. An SARC-F cut-off $\geq$4 had low to moderate sensitivity (47.4%) and high specificity (87.3%) to screen for poor functional outcomes (positive predictive value = 44.3%, negative predictive value = 88.6%, AUC = 0.702, 95% CI = 0.620–0.784).

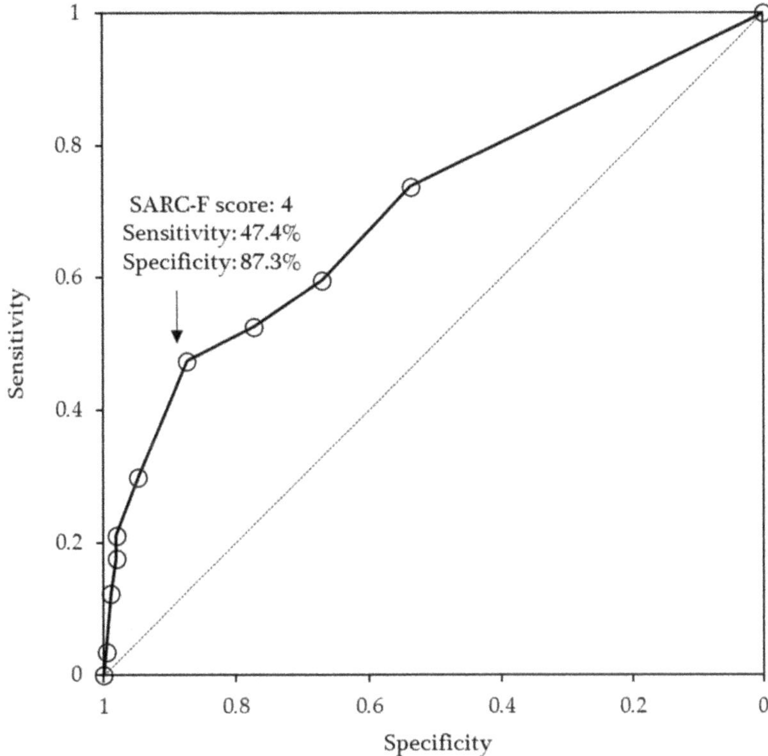

Figure 1. Receiver operating characteristic curves of SARC-F scores as an indicator of poor outcomes.

4. Discussion

This prospective cohort study examined the relationship between functional outcomes and pre-stroke SARC-F scores in stroke patients. The results obtained revealed the independent effects of pre-stroke SARC-F scores on functional outcomes and also supported the applicability of an SARC-F score ≥ 4 for predicting poor functional outcomes in stroke patients.

Previous studies reported a relationship between pre-stroke sarcopenia and poor outcomes [7–9]; however, these studies did not consider pre-stroke disability as a confounding factor. Sarcopenia is generally associated with disability [31–33], and pre-stroke disability has a negative impact on functional outcomes in stroke patients [29,34]. Therefore, the effects of pre-stroke sarcopenia on physical function need to be interpreted in consideration of pre-stroke disability. If a patient responds "unable", the results of SARC-F also need to be considered as the result of an assessment of disability. However, the present results demonstrated the independent effects of SARC-F scores on poor functional outcomes even after adjustments for pre-stroke disability. Therefore, pre-stroke SARC-F scores are useful not only for sarcopenia screening, but also for predicting poor functional outcomes in stroke patients.

The cut-off SARC-F score for sarcopenia screening is ≥ 4; however, a previous study demonstrated the low sensitivity and high specificity of this score for detecting sarcopenia [11,14]. We also investigated whether a cut-off SARC-F score ≥ 4 had low to moderate sensitivity and high specificity for the screening of poor functional outcomes in stroke patients. A SARC-F score ≥ 4 was previously associated with poor functional outcomes; however, this cut-off value was not validated [9]. The present results confirmed the validity of a SARC-F score ≥ 4 in stroke patients.

There are a number of limitations that need to be addressed. Younger patients or those with consciousness disorders, severe cognitive dysfunction, or aphasia were excluded from the present study. Since severe stroke patients were not enrolled, the present results may not be applicable to all stroke patients. Furthermore, the present study was conducted in a small, single-center setting and did not adjust for many confounding factors. Another limitation is that the impact of SARC-F scores on mortality or long-term outcomes was not examined. Therefore, further multicenter, large-scale, and long-term studies are needed to obtain more general and useful results for these patients.

5. Conclusions

Pre-stroke SARC-F scores were associated with poor functional outcomes in stroke patients even after adjustments for confounding factors, such as pre-stroke disability or a risk of malnutrition, and a SARC-F score ≥ 4 was suitable for predicting poor outcomes in these patients. Assessments of the pre-stroke status using SARC-F scores may be useful for predicting the outcomes of stroke patients.

Author Contributions: Conceptualization, M.N., H.K., M.Y. and M.K.; methodology, M.N.; formal analysis, M.N.; investigation, M.N., H.K., M.Y. and M.K.; data curation, M.N.; writing—original draft preparation, M.N.; writing—review and editing, H.K., M.Y. and M.K. All authors have read and agreed to the published version of the manuscript.

Funding: This work was supported by a research grant from JSPS KAKENHI (grant number 21K11182).

Institutional Review Board Statement: The present study was performed according to the guidelines of the Declaration of Helsinki and approved by the Ethics Committee of Konan Women's University (protocol code 2015020 and date of approval April 2016).

Informed Consent Statement: Informed consent was obtained from all subjects involved in the present study.

Data Availability Statement: The data presented in this study are available upon request from the corresponding author following permission by the Ethics Committee and the hospital at which the study was conducted.

Acknowledgments: We thank the individuals who participated in this study and the physical therapy staff of the Department of Rehabilitation, Itami Kousei Neurosurgical Hospital for their assistance.

Conflicts of Interest: The authors declare no conflict of interest.

References

1. Hardie, K.; Hankey, G.J.; Jamrozik, K.; Broadhurst, R.J.; Anderson, C. Ten-Year Risk of First Recurrent Stroke and Disability after First-Ever Stroke in the Perth Community Stroke Study. *Stroke* **2004**, *35*, 731–735. [CrossRef]
2. Luft, A.R.; Forrester, L.; Macko, R.F.; McCombe-Waller, S.; Whitall, J.; Villagra, F.; Hanley, D.F. Brain Activation of Lower Extremity Movement in Chronically Impaired Stroke Survivors. *NeuroImage* **2005**, *26*, 184–194. [CrossRef]
3. Ryan, A.S.; Ivey, F.M.; Serra, M.C.; Hartstein, J.; Hafer-Macko, C.E. Sarcopenia and Physical Function in Middle-Aged and Older Stroke Survivors. *Arch. Phys. Med. Rehabil.* **2017**, *98*, 495–499. [CrossRef]
4. Scherbakov, N.; von Haehling, S.; Anker, S.D.; Dirnagl, U.; Doehner, W. Stroke Induced Sarcopenia: Muscle Wasting and Disability after Stroke. *Int. J. Cardiol.* **2013**, *170*, 89–94. [CrossRef]
5. Yoshimura, Y.; Wakabayashi, H.; Bise, T.; Tanoue, M. Prevalence of Sarcopenia and Its Association with Activities of Daily Living and Dysphagia in Convalescent Rehabilitation Ward Inpatients. *Clin. Nutr.* **2018**, *37*, 2022–2028. [CrossRef]
6. Yoshimura, Y.; Wakabayashi, H.; Bise, T.; Nagano, F.; Shimazu, S.; Shiraishi, A.; Yamaga, M.; Koga, H. Sarcopenia is Associated with Worse Recovery of Physical Function and Dysphagia and a Lower Rate of Home Discharge in Japanese Hospitalized Adults Undergoing Convalescent Rehabilitation. *Nutrition* **2019**, *61*, 111–118. [CrossRef]
7. Ohyama, K.; Watanabe, M.; Nosaki, Y.; Hara, T.; Iwai, K.; Mokuno, K. Correlation Between Skeletal Muscle Mass Deficit and Poor Functional Outcome in Patients with Acute Ischemic Stroke. *J. Stroke Cerebrovasc. Dis.* **2020**, *29*, 104623. [CrossRef]
8. Abe, T.; Iwata, K.; Yoshimura, Y.; Shinoda, T.; Inagaki, Y.; Ohya, S.; Yamada, K.; Oyanagi, K.; Maekawa, Y.; Honda, A.; et al. Low Muscle Mass is Associated with Walking Function in Patients with Acute Ischemic Stroke. *J. Stroke Cereb. Dis.* **2020**, *29*, 105259. [CrossRef]
9. Nozoe, M.; Kanai, M.; Kubo, H.; Yamamoto, M.; Shimada, S.; Mase, K. Prestroke Sarcopenia and Functional Outcomes in Elderly Patients Who Have Had an Acute Stroke: A Prospective Cohort Study. *Nutrition* **2019**, *66*, 44–47. [CrossRef]

10. Malmstrom, T.K.; Miller, D.K.; Simonsick, E.M.; Ferrucci, L.; Morley, J.E. SARC-F: A Symptom Score To Predict Persons with Sarcopenia at Risk for Poor Functional Outcomes. *J. Cachexia-Sarcopenia Muscle* **2016**, *7*, 28–36. [CrossRef]
11. Ida, S.; Kaneko, R.; Murata, K. SARC-F for Screening of Sarcopenia among Older Adults: A Meta-Analysis of Screening Test Accuracy. *J. Am. Med. Dir. Assoc.* **2018**, *19*, 685–689. [CrossRef]
12. Tanaka, S.; Kamiya, K.; Hamazaki, N.; Matsuzawa, R.; Nozaki, K.; Maekawa, E.; Noda, C.; Yamaoka-Tojo, M.; Matsunaga, A.; Masuda, T.; et al. Utility of SARC-F for Assessing Physical Function in Elderly Patients with Cardiovascular Disease. *J. Am. Med. Dir. Assoc.* **2017**, *18*, 176–181. [CrossRef]
13. Chen, L.-K.; Woo, J.; Assantachai, P.; Auyeung, T.-W.; Chou, M.-Y.; Iijima, K.; Jang, H.C.; Kang, L.; Kim, M.; Kim, S.; et al. Asian Working Group for Sarcopenia: 2019 Consensus Update on Sarcopenia Diagnosis and Treatment. *J. Am. Med. Dir. Assoc.* **2020**, *21*, 300–307. [CrossRef]
14. Lu, J.-L.; Ding, L.-Y.; Xu, Q.; Zhu, S.-Q.; Xu, X.-Y.; Hua, H.-X.; Chen, L.; Xu, H. Screening Accuracy of SARC-F for Sarcopenia in the Elderly: A Diagnostic Meta-Analysis. *J. Nutr. Health Aging* **2021**, *25*, 172–182. [CrossRef]
15. Ishida, Y.; Maeda, K.; Nonogaki, T.; Shimizu, A.; Yamanaka, Y.; Matsuyama, R.; Kato, R.; Ueshima, J.; Murotani, K.; Mori, N. SARC-F as a Screening Tool for Sarcopenia and Possible Sarcopenia Proposed by AWGS 2019 in Hospitalized Older Adults. *J. Nutr. Health Aging* **2020**, *24*, 1053–1060. [CrossRef] [PubMed]
16. Kera, T.; Kawai, H.; Hirano, H.; Kojima, M.; Watanabe, Y.; Motokawa, K.; Fujiwara, Y.; Osuka, Y.; Kojima, N.; Kim, H.; et al. Limitations of SARC-F in the Diagnosis of Sarcopenia in Community-Dwelling Older Adults. *Arch. Gerontol. Geriatr.* **2020**, *87*, 103959. [CrossRef]
17. Erbas Sacar, D.; Kilic, C.; Karan, M.A.; Bahat, G. Ability of SARC-F to Find Probable Sarcopenia Cases in Older Adults. *J. Nutr. Health Aging* **2021**, *25*, 757–761. [CrossRef]
18. Bahat, G.; Ozkok, S.; Kilic, C.; Karan, M.A. SARC-F Questionnaire Detects Frailty in Older Adults. *J. Nutr. Health Aging.* **2021**, *25*, 448–453. [CrossRef]
19. Ishida, Y.; Maeda, K.; Ueshima, J.; Shimizu, A.; Nonogaki, T.; Kato, R.; Matsuyama, R.; Yamanaka, Y.; Mori, N. The SARC-F Score on Admission Predicts Falls during Hospitalization in Older Adults. *J. Nutr. Health Aging* **2021**, *25*, 399–404. [CrossRef]
20. Marini, A.C.B.; Perez, D.R.S.; Fleuri, J.A.; Pimentel, G.D. SARC-F Is Better Correlated with Muscle Function Indicators than Muscle Mass in Older Hemodialysis Patients. *J. Nutr. Health Aging* **2020**, *24*, 999–1002. [CrossRef]
21. Siqueira, J.M.; de Oliveira, I.C.; Soares, J.D.; Pimentel, G.D.; Pimentel, G.D. SARC-F Has Low Correlation and Reliability with Skeletal Muscle Mass Index in Older Gastrointestinal Cancer Patients. *Clin. Nutr.* **2021**, *40*, 890–894. [CrossRef]
22. Kang, M.K.; Kim, T.J.; Kim, Y.; Nam, K.-W.; Jeong, H.-Y.; Kim, S.K.; Lee, J.S.; Ko, S.-B.; Yoon, B.-W. Geriatric Nutritional Risk Index Predicts Poor Outcomes in Patients with Acute Ischemic Stroke-Automated Undernutrition Screen Tool. *PLoS ONE* **2020**, *15*, e0228738. [CrossRef]
23. Fearon, P.; McArthur, K.S.; Garrity, K.; Graham, L.J.; McGroarty, G.; Vincent, S.; Quinn, T.J. Prestroke Modified Rankin Stroke Scale Has Moderate Interobserver Reliability and Validity in an Acute Stroke Setting. *Stroke* **2012**, *43*, 3184–3188. [CrossRef]
24. Weisscher, N.; Vermeulen, M.; Roos, Y.B.; de Haan, R.J. What Should Be Defined as Good Outcome in Stroke Trials; A Modified Rankin Score of 0–1 or 0–2? *J. Neurol.* **2008**, *255*, 867–874. [CrossRef]
25. Rangaraju, S.; Haussen, D.C.; Nogueira, R.G.; Nahab, F.; Frankel, M. Comparison of 3-Month Stroke Disability and Quality of Life across Modified Rankin Scale Categories. *Interv. Neurol.* **2017**, *6*, 36–41. [CrossRef]
26. Nakayama, H.; Jørgensen, H.S.; O Raaschou, H.; Olsen, T.S. The Influence of Age on Stroke Outcome. The Copenhagen Stroke Study. *Stroke* **1994**, *25*, 808–813. [CrossRef]
27. Carcel, C.; Wang, X.; Sandset, E.C.; Delcourt, C.; Arima, H.; Lindley, R.; Hackett, M.L.; Lavados, P.; Robinson, T.G.; Muñoz Venturelli, P.; et al. Sex Differences in Treatment and Outcome after Stroke: Pooled Analysis Including 19,000 Participants. *Neurology* **2019**, *93*, 2170–2180. [CrossRef]
28. Sato, S.; Toyoda, K.; Uehara, T.; Toratani, N.; Yokota, C.; Moriwaki, H.; Naritomi, H.; Minematsu, K. Baseline NIH Stroke Scale Score Predicting Outcome in Anterior and Posterior Circulation Strokes. *Neurology* **2008**, *70*, 2371–2377. [CrossRef]
29. Ganesh, A.; Luengo-Fernandez, R.; Pendlebury, S.T.; Rothwell, P.M. Long-Term Consequences of Worsened Poststroke Status in Patients with Premorbid Disability. *Stroke* **2018**, *49*, 2430–2436. [CrossRef] [PubMed]
30. Davis, J.P.; Wong, A.A.; Schluter, P.J.; Henderson, R.D.; O'Sullivan, J.D.; Read, S.J. Impact of Premorbid Undernutrition on Outcome in Stroke Patients. *Stroke* **2004**, *35*, 1930–1934. [CrossRef]
31. Phillips, A.; Strobl, R.; Vogt, S.; Ladwig, K.-H.; Thorand, B.; Grill, E. Sarcopenia Is Associated with Disability Status-Results from the KORA-Age Study. *Osteoporos. Int.* **2017**, *28*, 2069–2079. [CrossRef] [PubMed]
32. Fang, Q.; Zhu, G.; Huang, J.; Pan, S.; Fang, M.; Li, Q.; Yin, Q.; Liu, X.; Tang, Q.; Huang, D.; et al. Current Status of Sarcopenia in the Disabled Elderly of Chinese Communities in Shanghai: Based on the Updated EWGSOP Consensus for Sarcopenia. *Front. Med.* **2020**, *12*, 552415. [CrossRef] [PubMed]
33. Shimada, H.; Tsutsumimoto, K.; Doi, T.; Lee, S.; Bae, S.; Nakakubo, S.; Makino, K.; Arai, H. Effect of Sarcopenia Status on Disability Incidence Among Japanese Older Adults. *J. Am. Med. Dir. Assoc.* **2021**, *22*, 846–852. [CrossRef]
34. Han, T.S.; Fry, C.H.; Gulli, G.; Affley, B.; Robin, J.; Irvin-Sellers, M.; Fluck, D.; Kakar, P.; Sharma, S.; Sharma, P. Prestroke Disability Predicts Adverse Poststroke Outcome: A Registry-Based Prospective Cohort Study of Acute Stroke. *Stroke* **2020**, *51*, 594–600. [CrossRef]

Article

Stored Energy Increases Body Weight and Skeletal Muscle Mass in Older, Underweight Patients after Stroke

Yoshihiro Yoshimura [1,*], Hidetaka Wakabayashi [2], Ryo Momosaki [3], Fumihiko Nagano [1], Takahiro Bise [1], Sayuri Shimazu [1] and Ai Shiraishi [1]

[1] Center for Sarcopenia and Malnutrition Research, Kumamoto Rehabilitation Hospital, Kumamoto 869-1106, Japan; f-nagano@kumareha.jp (F.N.); asian.dub.foundation00@gmail.com (T.B.); shimazu@kumareha.jp (S.S.); ai.shiraishi0913@gmail.com (A.S.)
[2] Department of Rehabilitation Medicine, Tokyo Women's Medical University Hospital, Tokyo 162-0054, Japan; noventurenoglory@gmail.com
[3] Department of Rehabilitation Medicine, Mie University Graduate School of Medicine, Tsu 514-8507, Japan; momosakiryo@gmail.com
* Correspondence: hanley.belfus@gmail.com; Tel.: +81-96-232-3111

Abstract: We conducted a retrospective observational study in 170 older, underweight patients after stroke to elucidate whether stored energy was associated with gains in body weight (BW) and skeletal muscle mass (SMM). Energy intake was recorded on admission. The energy requirement was estimated as actual BW (kg) × 30 (kcal/day), and the stored energy was defined as the energy intake minus the energy requirement. Body composition was measured by bioelectrical impedance analysis. The study participants gained an average of 1.0 ± 2.6 kg of BW over a mean hospital stay of 100 ± 42 days with a mean stored energy of 96.2 ± 91.4 kcal per day. They also gained an average of 0.2 ± 1.6 kg of SMM and 0.5 ± 2.3 kg of fat mass (FM). This means about 9600 kcal were needed to gain 1 kg of BW. In addition, a 1 kg increase in body weight resulted in a 23.7% increase in SMM and a 45.8% increase in FM. Multivariate regression analyses showed that the stored energy was significantly associated with gains in BW and SMM. Aggressive nutrition therapy is important for improving nutritional status and function in patients with malnutrition and sarcopenia.

Keywords: stored energy; body weight gain; skeletal muscle mass gain; malnutrition; aggressive rehabilitation nutrition

Citation: Yoshimura, Y.; Wakabayashi, H.; Momosaki, R.; Nagano, F.; Bise, T.; Shimazu, S.; Shiraishi, A. Stored Energy Increases Body Weight and Skeletal Muscle Mass in Older, Underweight Patients after Stroke. *Nutrients* **2021**, *13*, 3274. https://doi.org/10.3390/nu13093274

Academic Editor: Susanna C. Larsson

Received: 25 June 2021
Accepted: 17 September 2021
Published: 19 September 2021

Publisher's Note: MDPI stays neutral with regard to jurisdictional claims in published maps and institutional affiliations.

1. Introduction

Malnutrition is commonly observed in the geriatric population and is associated with adverse outcomes in geriatric rehabilitation patients. Stroke or hip fracture patients are often transferred from acute-care hospital wards to convalescent rehabilitation wards, and these patient groups show a high prevalence of malnutrition and sarcopenia at 49–67% [1,2] and 40–53% [3,4], respectively. Malnutrition, weight loss, body mass index (BMI) lower than 20 kg/m^2, sarcopenia, and reduced nutritional intake are established independent factors that negatively influence functional recovery in older inpatients [5–7]. The goal of geriatric rehabilitation is to promote functional recovery and thereby allow hospitalized patients to return to their homes. Therefore, it is important to improve the nutritional status and sarcopenia in older rehabilitation patients to maximize favorable outcomes.

Aggressive nutrition therapy improves the outcomes of older hospitalized patients [8–10]. Weight gain and muscle mass gain during hospitalization have a positive effect on activities of daily living (ADL) in these patients [11,12]. In older patients with low BMI or sarcopenia, daily energy expenditure and energy stores need to be considered in terms of energy needs. Energy expenditure in healthy individuals consists of resting metabolic rate, diet-induced thermogenesis, and energy expenditure related to activity and illness. On the other hand, for patients who are malnourished and for those with sarcopenia, energy requirements need to be estimated by adding stored energy for the restoration of lean body mass [8]. To gain 1 kg of body weight (BW), individuals aged 10–40 years need 7500 kcal [13,14].

However, there is a lack of evidence on the relationship between stored energy and weight gain and skeletal muscle mass (SMM) gain in geriatric rehabilitation patients. Hence, we conducted a retrospective cohort study to determine whether the stored energy intake was associated with gains in BW and SMM, as well as how much energy was needed to gain 1 kg of BW in older patients with low BW undergoing convalescent rehabilitation after stroke.

2. Materials and Methods

2.1. Study Design and Participants

We conducted a retrospective cohort study at a post-acute care hospital, which included convalescent rehabilitation wards with a total of 135 beds [15]. The study was conducted between January 2016 and December 2020 and included all stroke patients newly admitted to the wards who were over 70 years old and had a BMI of less than 20.0 kg/m^2 [16]. Patients were excluded from this study if they refused to participate, had incomplete data, or had edema of the extremities, pleural or ascites fluid, or altered consciousness. The observation period lasted from the date of admission to the date of discharge.

2.2. Data Collection

The baseline patient demographic characteristics were recorded upon admission, including age, sex, body mass index, stroke type, stroke history, days from stroke onset to admission [17]; nutritional status, assessed using the Mini Nutritional Assessment-Short Form (MNA-SF) [18]; swallowing status, assessed using the Food Intake Level Scale (FILS) and presence of dysphagia requiring supplemental feeding defined by the FILS scores <7 [19]; comorbidities, assessed using the Charlson Comorbidity Index (CCI) [20]; and premorbid ADL, assessed using the modified Rankin Scale (mRS) [21]. Information was collected on the presence of paralysis and localization and stage of paralysis according to the Brunnstrom recovery stages (BRS) [22]. Blood sampling data on admission, including albumin, hemoglobin, and C-reactive protein levels were recorded. The number of drugs prescribed at the time of admission was recorded. Data on the total rehabilitation therapy received during hospitalization (units per day, 1 unit = 20 min of therapy) were extracted from medical records.

Within 72 h of admission, bioelectrical impedance analysis (BIA) data of the SMM and fat mass (FM), hand grip strength (HG), and functional independence measure (FIM) scores for physical and cognitive function (FIM-motor and FIM-cognitive) [23] were obtained. FIM gain was calculated by subtracting the FIM score at discharge from the FIM score at admission. The BIA measurements were carried out using the latest version of a validated instrument (InBody S10: InBody, Tokyo, Japan) and following a standard protocol explained elsewhere [24]. SMM and FM each were divided by the square of height (m) and indexed as the skeletal muscle mass index (SMI) and fat mass index (FMI). HG was measured using the Smedley hand dynamometer (TTM, Tokyo, Japan) in the non-dominant hand (or in case of hemiparesis, in the non-paralyzed hand), with the patient in the standing or seated position (depending on their ability) and with their arms relaxed at their side; measurements were taken three times, and the highest value was recorded. Sarcopenia was diagnosed when both the SMI and HG values were low, in accordance with the Asian Working Group for Sarcopenia criteria 2019 [25].

2.3. Stored Energy and Nutritional Intakes

Energy and protein intakes were calculated by a nurse or dietitian who visually assessed the ratio of intake to the amount provided to the patient. The intake from three servings each of breakfast, lunch, and dinner (a total of nine servings) was recorded [26], and the average of each value divided by three was used as the daily intake. In the case of enteral (EN) and parenteral nutrition (PN), the energy and protein doses for the first 72 h of hospitalization were recorded, and the respective values divided by 3 were used as daily intake. If oral intake was combined with EN or PN, the respective energy and protein intake (administered) were added up. In addition, nutrient intake was calculated by dividing each intake by the actual body weight on admission. Nutritional intake was recorded at admission and at discharge.

The energy requirement for older patients undergoing stroke rehabilitation was estimated as actual BW (kg) $\times$ 30 (kcal/day) [8,27], and the stored energy was defined as the energy intake minus the energy requirement. According to the stored energy on admission, patients were divided into two groups: those whose stored energy was greater than 0 kcal/day and those whose stored energy was less than 0 kcal/day.

2.4. Outcomes

The primary outcome was BW gain, which was defined as the change in BW during hospitalization (BW at discharge—BW at admission). The secondary outcome was SMM gain, which was defined as the change in SMM during hospitalization (SMM at discharge—SMM at admission).

2.5. Convalescent Rehabilitation

The convalescent rehabilitation program (up to 3 h per day) was performed according to the guidelines of the National Health Insurance System. The program was tailored to suit the functional abilities and disabilities of the patient, such as physical therapy including paralyzed limb facilitation (for leg paralysis), range of motion exercises, basic movement training (mainly for the legs), walking training, resistance training (such as chair-stand exercises [28]), and ADL training [29].

For nutritional management, nutritional screening and nutritional assessment were conducted for all patients, and under the guidance of the dietitians and nutrition support team, active nutritional support was provided, including high-energy and/or high-protein meals. In addition, nutrition management was tailored to each patient's condition and nutritional needs by adjusting energy and protein according to changes in rehabilitation time and load [30].

Dysphagia rehabilitation was customized to the patients' swallowing abilities and function, and included oral management and exercise, indirect (without food) and direct (with food) exercises, and diet modification through multi-occupational collaboration with speech and swallowing therapists, dental hygienists, and ward staff [31].

Oral management included oral screening, assessment, education, counseling, treatment (oral care), oral and dysphagia rehabilitation, medical treatment by a dentist, and practicing in cooperation with a multidisciplinary team [31]. Ward dental hygienists conducted oral and dysphagia rehabilitation, including indirect and direct (oral intake) exercises at the patient's bedside [32].

Medication management was carried out by multidisciplinary teams, including pharmacists. Pharmacotherapy is one of the factors that affect the nutritional state of older people. Polypharmacy and inappropriate medications were corrected, and medications that could affect nutritional status were managed throughout the hospital stay [33].

2.6. Sample Size Calculation

The sample size for statistical power was calculated by using data from our previous study [34], and the results showed that the BW of patients at hospital admission was normally distributed with a standard deviation (SD) of 8.0. If the true difference between the mean values of BW at discharge of patients with more and less stored energy is 4.0, a sample size of at least 64 participants would be required in each group to reject the null hypothesis with a power of 0.8 and an alpha error of 0.05, and this would support the validity of our results.

2.7. Statistical Analysis

The values were reported as the mean (SD) for parametric data or as the median (interquartile range; IQR) and numbers (%) for non-parametric and categorical data, respectively. In the univariate analyses, the patients were stratified according to the intake of stored energy (with or without stored energy). Comparisons between groups were made using the t-test, Mann-Whitney *U* test, and chi-square test, as suitable.

Multiple linear regression analysis was carried out to determine whether the stored energy was independently associated with the study outcomes, including BW gain and SMM gain during hospitalization. The covariates of age, sex, length of hospital stay, BRS of the lower limb, FIM-motor and FIM-cognitive scores at admission, total rehabilitation therapy (units/day), energy intake at baseline, FMI and SMI on admission, and stroke type, all of which are considered to be confounding factors affecting BW gain, were included in the model. Multicollinearity was assessed using the variance inflation factor (VIF): a VIF value of 1–10 indicated the absence of multicollinearity. Statistical significance was set at *p*-values of <0.05. All analyses were conducted using IBM SPSS version 21 (Armonk, NY, USA).

2.8. Ethics

This study was approved by the Institutional Review Board (approval no.: 169-210617) of the hospital where the study was conducted. Written informed consent could not be obtained because of the constraints imposed by the retrospective study design, although the participants could withdraw from this study at any time by using an opt-out procedure. This study was conducted in accordance with the Declaration of Helsinki and ethical guidelines for medical and health research involving human subjects.

3. Results

During the study period, a total of 843 stroke patients were newly admitted to the hospital. Patients with missing data (*n* = 22) and altered consciousness (*n* = 8) were excluded. Data from 813 patients were considered for inclusion in the study. Among them, patients aged 70 years or younger (*n* = 418) and those with BMI $\geq$ 20 (*n* = 583) were excluded. Finally, 170 patients who were over 70 years of age and had a BMI of <20.0 were included in the analysis (Figure 1).

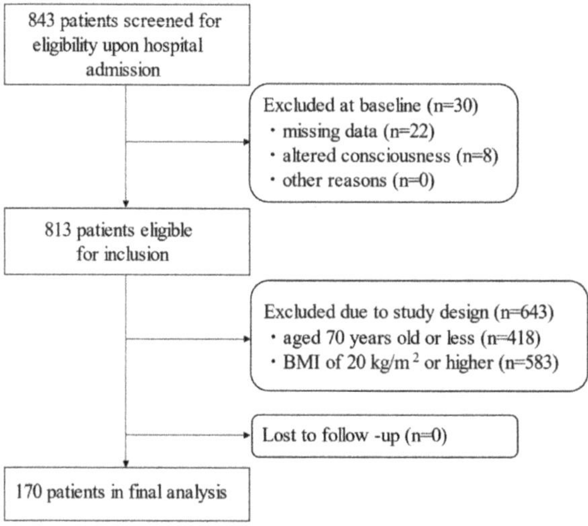

Figure 1. Flowchart of participant screening, inclusion criteria, and follow-up.

The patients' baseline characteristics are presented in Table 1. The mean age was 83.1 ± 6.2 years; 52% of the patients were male. The median BW and SMM were 41.6 kg (IQR: 38.2–44.1) and 10.9 kg (IQR: 8.8–13.2), respectively. The median MNA-SF was 4 (IQR: 3–6), suggesting that a large population of the patients was malnourished. The median FIM-motor and FIM-cognitive scores were 32 (IQR: 14–58) and 17 (IQR: 10–25), respectively, suggesting that a large proportion of patients was physically dependent at the baseline. Sarcopenia was observed in 84.7% of the patients (n = 144), with a median HG of 12.5 kg (IQR: 5.6–17.8) and SMI of 4.9 kg/m^2 (IQR: 4.0–5.7).

Table 1. Baseline patient characteristics and between-group characteristics of patients with and without stored energy intake.

	Total (n = 170)	Patient without Stored Energy (n = 56)	Patient with Stored Energy (n = 114)	p
Age, y, mean (SD)	83.1 (6.2)	84.9 (6.4)	82.1 (5.9)	0.005
Sex (male), n (%)	52 (30.6)	22 (39.3)	30 (26.3)	0.111
Stroke type, n (%)				
Cerebral infarction	114 (67.1)	46 (82.1)	68 (59.6)	0.506
Cerebral hemorrhage	42 (24.7)	8 (14.3)	34 (29.8)	0.037
SAH	12 (7.1)	0 (0.0)	12 (10.5)	0.009
Stroke history, n (%)	46 (27.1)	18 (32.1)	28 (24.6)	0.359
Pre-stroke mRS, score, median (IQR)	1 (0,3)	0 (0,3]	1 (0,2)	0.876
Onset-admission days	12 (9,19)	12 (10,15)	14 (9,24)	0.291
Paralysis, n (%)				
Right/Left/Both	64 (37.6)/76 (44.7)/8 (4.7)	22 (39.3)/26 (46.4)/2 (3.6)	42 (36.8)/50 (43.9)/6 (5.3)	0.866
BRS, median (IQR)	5 (2,6)	3 (1,5)	5 (3,5)	
Upper limb	5 (2,6)	2 (1,5)	5 (3,5)	0.001
Hand-finger	5 (2,6)	3 (1,5)	5 (3,5)	<0.001
Lower limb				0.014
FIM, *score*, median (IQR)				
Total	49 (25,84)	39 (22,69)	59 (26,86)	0.034
Motor	32 (14,58)	20 (13,53)	37 (15,59)	0.033
Cognitive	17 (10,25)	16 (8,25)	18 (11,26)	0.232
Swallowing status				
FILS, score, median (IQR)	7 (6,9)	7 (2,7)	7 (7,9)	0.004
Dysphagia, n (%)	44 (25.9)	22 (39.3)	22 (19.3)	0.009
CCI, median (IQR)	3 (2,4)	3 (2,4)	3 (2,4)	0.687
MNA-SF, median (IQR)	4 (3,6)	3 (2,6)	4 (3,6)	0.047
Body composition, median (IQR]				
BW, kg	41.0 (38.2, 44.1)	43.6 (40.9, 47.5)	39.8 (38.0, 42.2)	<0.001
BMI, kg/m^2	18.3 (16.7, 19.3)	18.4 (17.4, 19.4)	18.3 (16.5, 19.2)	0.268
SMM, kg	10.9 (8.8, 13.2)	12.6 (9.4, 15.1)	10.1 (8.8, 12.2)	<0.001
SMI, kg/m^2	4.9 (4.0, 5.7)	5.2 (4.1, 6.1)	4.6 (4.0, 5.3)	0.002
FM, kg	9.8 (8.0, 13.5)	11.1 (8.3, 15.1)	9.5 (7.2, 12.9)	0.021
FMI, kg/m^2	4.3 (3.3, 6.2)	4.4 (3.3, 6.5)	4.2 (3.2, 6.0)	0.295
HG, kg, median (IQR)	12.5 (5.6, 17.8)	14.4 (4.5, 20.6)	12.3 (6.0, 15.1)	0.199
Sarcopenia, n (%)	144 (84.7)	46 (82.1)	98 (86.0)	0.506
Total drug number, median (IQR)	5 (3, 8)	6 (4, 8)	5 (3, 7)	0.032
Laboratory data, mean (SD)				
Albumin, g/dL	3.28 (0.46)	3.3 (0.3)	3.3 (0.4)	0.503
Hemoglobin, g/dL	12.12 (1.40)	12.3 (1.4)	12.0 (1.3)	0.167
C-reactive protein, mg/dL	1.8 (2.7)	1.9 (2.4)	1.6 (2.9)	0.447

BMI, body mass index; BRS, Brunnstrom Recovery Stage; BW, body weight; CCI, Charlson's Comorbidity Index; FILS, Food Intake Level Scale; FIM, Functional Independence Measure; FM, fat mass; FMI, fat mass index; HG, handgrip strength; MNA-SF, Mini Nutritional Assessment-Short Form; mRS, modified Rankin Scale; SAH, subarachnoid hemorrhage; SMI, skeletal muscle mass index; SMM, skeletal muscle mass.

Overall, the study patients gained an average of 1.0 ± 2.6 kg of BW over a mean hospital stay of 100.2 ± 42.3 days, with a mean stored energy intake of 96.2 ± 91.4 kcal per day. They also gained an average of 0.2 ± 1.6 kg of SMM and 0.5 ± 2.3 kg of FM, respectively. This means that it took about 9600 kcal (9614.2) for the patient to gain 1 kg of BW (mean stored energy per day [96.2] × mean number of days in hospital [100.2]/mean BW increase [1.0]). In addition, a 1 kg increase in body weight resulted in a 23.7% increase in SMM (mean SMM increase ÷ mean BW increase × 100) and a 45.8% increase in FM (mean FM increase ÷ mean BW increase × 100). Table 2 shows a two-group comparison of the nutritional profiles of patients with and without stored energy intake by stroke type. Except for patients with subarachnoid hemorrhage (SAH), patients with stored energy intake of more than 0 kcal/day consumed more energy and protein on admission and at discharge than those with stored energy intake of less than 0 kcal/day. All patients with SAH were supplemented with stored energy. In the univariate analysis, there was no difference in BW gain or SMM gain between the two groups.

Table 2. Univariate analyses for outcomes between groups of patients with and without stored energy intake by stroke type.

	Cerebral Infarction			Cerebral Hemorrhage			Subarachnoid Hemorrhage		
	without Stored Energy (*n* = 46)	with Stored Energy (*n* = 68)	*p*	without Stored Energy (*n* = 8)	with Stored Energy (*n* = 34)	*p*	without Stored Energy (*n* = 0)	with Stored Energy (*n* = 12)	*p*
Energy intake									
On admission, kcal/kg/day	28.0 (26.0, 29.7)	35.3 (33.4, 37.9)	<0.001	28.4 (25.9, 29.0)	35.0 (31.5, 40.6)	<0.001	–	31.6 (31.1, 39.2)	–
On admission, kcal/day	1200 (1200, 1200)	1400 (1400, 1400)	<0.001	1350 (1125, 1525)	1400 (1200, 1600)	0.426	–	1300 (1200, 1500)	–
Set energy, kcal/kg/day	1299 (1208, 1410)	1205 (1101, 1263)	<0.001	1449 (1323, 1575)	1170 (1140, 1323)	0.005	–	1158 (1150, 1240)	–
Stored energy, kcal/day	−126 (−214, −64)	195 (140, 299)		−76 (176, −50)	200 (60, 308)		–	67 (46, 262)	–
On discharge, kcal/kg/day	31.4 (26.7, 38.3)	36.7 (35.4, 40.8)	<0.001	30.7(27.4, 34.4)	39.0 (34.7, 42.4)	0.001	–	32.8 (29.5, 38.4)	–
Protein intake g/kg/day									
On admission	1.1 (1.0, 1.2)	1.3 (1.2, 1.5)	<0.001	1.2 (1.1, 1.3)	1.3 (1.1, 1.5)	0.158	–	1.3 (1.1, 1.6)	–
On discharge	1.1 (1.0, 1.1)	1.3 (1.1, 1.4)	<0.001	1.2 (1.1, 1.2)	1.2 (1.1, 1.4)	0.001	–	1.2 (1.0, 1.6)	–
Change in BC kg									
BW	0.6 (0.0, 2.8)	0.3 (−0.9, 2.5)	0.101	−0.3 (−1.2, 0.7)	1.3 (−0.1, 3.0)	0.123	–	1.6 (0.5, 2.9)	–
SMM	0.3 (−0.1, 0.5)	0.3 (−0.4, 0.7)	0.947	−0.1 (−0.4, 0.0)	0.3 (0.1, 1.5)	0.002	–	0.7 (0.3, 1.8)	–
FM	1.1 (−1.2, 1.9)	0.5 (−1.0, 1.7)	0.721	−0.5 (−1.4, −0.0)	0.2 (−1.0, 0.8)	0.229	–	1.3 (−0.5, 2.8)	–
Changes in HG kg	0.1 (−1.5, 3.1)	1.3 (0.0, 5.1)	0.146	3.6 (−1.8, 4.3)	3.5 (0.0, 7.5)	0.503	–	2.2 (0.5, 6.3)	–
FIM gain									
total	21 (8, 34)	23 (16, 37)	0.337	38 (16, 57)	31 (5, 53)	0.247	–	27 (1, 29)	–
motor	19 (3, 29)	19 (12, 27)	0.495	30 (13, 48)	29 (3, 42)	0.248	–	19 (1, 23)	–
cognitive	3 (1, 6)	5 (2, 7)	0.182	5 (3, 7)	4 (0, 7)	0.244	–	6 (1, 9)	–
FILS on discharge	8 (7, 10)	9 (7, 10)	0.418	9 (8, 9)	9 (7, 10)	0.84	–	8 (5, 9)	–
Length of stay, day	94 (68, 134)	83 (57, 124)	0.636	148 (136, 159)	91 (76, 118)	0.001	–	114 (58, 174)	–
Rehabilitation units/day	8 (7, 8)	8 (7, 8)	0.331	8 (8, 8)	8 (7, 8)	0.129	–	7 (5, 8)	–

BC, body composition; BW, body weight; FILS, Food Intake Level Scale; FIM, Functional Independence Measure; FM, fat mass; HG, handgrip strength; SMM, skeletal muscle mass.

In the multivariate linear regression analysis, the stored energy intake was significantly and positively associated with BW gain during hospitalization ($\beta = 0.256$, $p = 0.005$) after adjusting for potential confounders, including age, sex, baseline energy intake, FMI and SMI (Table 3). Moreover, the stored energy intake was significantly and positively associated with SMM gain during hospitalization ($\beta = 0.263$, $p = 0.011$) in the same analysis model (Table 4).

Table 3. Multivariate regression analysis for BW gain during hospitalization.

	B (95% CI)	SE	β	*p*
Age, y	−0.036 (−0.109, 0.036)	0.037	−0.091	0.326
Sex (male)	0.466 (−0.508, 1.440)	0.493	0.087	0.346
Length of stay, day	0.011 (0.000, 0.023)	0.006	0.193	0.052
BRS-Lower limb	−0.160 (−0.406, 0.085)	0.124	−0.130	0.199
FIM-motor on admission	−0.021 (−0.055, 0.014)	0.017	−0.179	0.233
FIM-cognitive on admission	−0.029 (−0.094, 0.036)	0.033	−0.101	0.382
HG on admission, kg	0.132 (0.063, 0.201)	0.035	0.452	<0.001
Rehabilitation, units/day	−0.526 (−0.848, −0.204)	0.163	−0.275	0.002
Stored energy, kcal/day	0.003 (0.001, 0.005)	0.001	0.256	0.005
FMI on admission, kg/cm^2	0.298 (0.069, 0.527)	0.116	0.227	0.011
SMI on admission, kg/cm^2	−0.142 (−0.628, 0.344)	0.246	−0.064	0.564
Stroke type				
Cerebral infarction	0.035 (−1.387, 1.457)	0.72	0.007	0.961
Cerebral hemorrhage	−0.115 (−1.676, 1.445)	0.79	−0.020	0.884
SAH (control)	–	–	–	–

BRS, Brunnstrom Recovery Stage; BW, body weight; FIM, Functional Independence Measure; FMI, fat mass index; HG, handgrip strength; SAH, subarachnoid hemorrhage; SMI, skeletal muscle mass index.

Table 4. Multivariate regression analysis for SMM gain during hospitalization.

	B (95% CI)	SE	β	*p* Value
Age, y	−0.045 (−0.105, 0.015)	0.03	−0.160	0.136
Sex (male)	0.576 (−0.056, 1.208)	0.318	0.166	0.074
Length of stay, day	−0.002 (−0.010, 0.007)	0.004	−0.047	0.676
BRS-Lower limb	0.074 (−0.109, 0.258)	0.092	0.09	0.424
FIM-motor on admission	−0.008 (−0.035, 0.019)	0.014	−0.111	0.55
FIM-cognitive on admission	−0.004 (−0.053, 0.046)	0.025	−0.020	0.881
HG on admission, kg	0.095 (0.036, 0.155)	0.03	0.478	0.002
Rehabilitation, units/day	−0.196 (−0.405, 0.014)	0.105	−0.168	0.066
Stored energy, kcal/day	0.002 (0.000, 0.004)	0.001	0.263	0.011
FMI on admission, kg/cm^2	0.034 (−0.141, 0.209)	0.088	0.038	0.702
SMI on admission, kg/cm^2	−1.275 (−1.628, −0.922)	0.178	−0.871	<0.001
Stroke type				
Cerebral infarction	−0.518 (−1.485, 0.448)	0.487	−0.150	0.29
Cerebral hemorrhage	−0.347 (−1.444, 0.749)	0.552	−0.090	0.531
SAH (control)	−	−	−	−

BRS, Brunnstrom Recovery Stage; BW, body weight; FIM, Functional Independence Measure; FMI, fat mass index; HG, handgrip strength; SAH, subarachnoid hemorrhage; SMI, skeletal muscle mass index.

4. Discussion

In this study, we determined whether stored energy intake was associated with gains in BW and SMM during hospitalization in older, underweight, post-stroke rehabilitation patients, and highlight two important findings: (1) Stored energy intake was associated with gain in BW and SMM, and (2) it took about 9600 kcal to gain 1 kg of body weight in these patients.

Stored energy intake was associated with gain in weight and muscle mass during hospitalization in geriatric rehabilitation after stroke. The results of the univariate analysis showed differences by stroke type, but the small sample size of 12 patients with SAH might have affected the results. On the other hand, since we adjusted for stroke type in the multivariate analysis, we consider that the effect of stroke type on the results was controlled. Considering the fact that weight gain and increased muscle mass have a positive impact on the ADLs of these patients [11,12], our findings suggest the need for enhanced nutritional management for older rehabilitation patients with low body weight. However, strong evidence exists that a large proportion of stroke patients do not even meet their estimated energy requirement both in hospital and after discharge [35]. An increase in energy demand related to rehabilitation, chronic diseases, and improvement of sarcopenia and malnutrition would likely result in negative energy balance in geriatric rehabilitation. Therefore, it is important to ensure the provision of individualized stored energy for weight gain in addition to energy requirements, while energy calculations should take into account post-stroke paralysis, muscle spasticity, physical disabilities, nutritional status, and age and sex in geriatric rehabilitation after stroke. Furthermore, attention also needs to be paid to medications that are often prescribed to post-stroke patients. For example, statins are myotoxic and have been suggested to be associated with the progression of sarcopenia [36]. On the other hand, there is evidence that statin use reduces the incidence of sarcopenia, so if a patient is a good candidate for statin use, it is necessary to take care of myopathy and sarcopenia while using statins [37].

To increase 1 kg of BW, about 9600 kcal were required for older rehabilitation patients who were underweight. Positive energy balance is necessary for growth, wound healing, and muscle gain, but increased intake of energy or prolonged intake can lead to overweight and obesity [14]. Therefore, specific numerical evidence is needed for nutritional planning so as to achieve a positive energy balance that can promote weight gain in underweight patients. In the literature, to gain 1 kg of lean body mass, individuals aged 10–40 years needed 7500 kcal [13,14], while older individuals needed 8800–22,600 kcal [38]. The diversity of energy requirements for weight gain probably depends on the subject's background, including age; sex; nutritional status; changes in body composition; the type, intensity, and duration of rehabilitation and exercise; presence of systemic inflammation; and co-morbidities. In this study, we estimated the stored energy for weight gain in lower-weight older patients undergoing stroke rehabilitation. We expect this estimate to help improve the nutritional status of older patients in daily clinical practice. However, further studies are needed to validate our result.

Ensuring that underweight older patients obtain the required level of stored energy is a practical nutritional management. This may also be true for underweight older patients in general, not just stroke patients. Aggressive rehabilitation nutrition improves nutritional and functional outcomes in patients with malnutrition and sarcopenia [8]. Malnutrition and sarcopenia negatively affect functional recovery and ADL, while nutrition improvement, including weight gain and muscle mass gain, is positively associated with functional recovery [11,12]. Aggressive nutrition therapy described here is characterized by the setting of energy requirements based on the amount energy expenditure per day plus stored energy. This is not the same as enhanced nutrition therapy and instead focuses on improving malnutrition and sarcopenia. Moreover, rehabilitation nutrition care process suggests that aggressive nutrition therapy should be performed in combination with aggressive exercise therapy [39]. Medication is another factor that can affect the nutritional status in older patients undergoing rehabilitation. Since some of the most important drugs that affect nutritional status are those that cause anorexia, rehabilitation pharmacotherapy should be practiced [33]. Therefore, it is necessary to reconcile the rehabilitation goal setting, the content, quantity, and quality of physical activity and exercise therapy, and the patient's general condition to set nutrition goals through multidisciplinary collaboration.

This study had some limitations. First, the daily stored energy intake was based on estimates at the time of admission and not on weight change over the course of the hospital stay. This could be a major limitation of this study, as energy adjustments are made as appropriate for weight changes over time in daily clinical practice. Moreover, determination of energy requirements by indirect calorimetry over time would provide more accurate estimates than those used in this study. Further, this study involved a single rehabilitation hospital in Japan, possibly limiting the generalizability of the findings. In addition, there was a selection bias in that the subjects were older hospitalized patients with low body weight. Future multicenter studies are needed to determine whether similar results can be obtained in diverse populations, especially by stroke type. Lastly, due to its retrospective study design, we were unable to obtain detailed information on nutritional therapy during hospitalization, possibly affecting the results. High-quality prospective studies adjusted for confounders are needed in the future.

5. Conclusions

Stored energy intake was associated with weight gain and muscle mass gain in underweight older patients undergoing stroke rehabilitation. Further, it took about 9600 kcal to gain 1 kg of body weight in these patients. Aggressive nutrition therapy is important for improving nutritional status and function in patients with malnutrition and sarcopenia.

Author Contributions: Conceptualization, Y.Y., H.W., R.M., F.N., T.B., S.S., and A.S.; data curation, Y.Y., F.N., T.B., S.S., and A.S.; formal analysis, Y.Y., H.W., R.M., F.N., T.B., S.S., and A.S.; writing—original draft preparation and writing—review and editing, Y.Y., H.W., R.M., F.N., T.B., S.S., and A.S. All authors have read and agreed to the published version of the manuscript.

Funding: This research received no external funding.

Institutional Review Board Statement: This study was approved by the Institutional Review Board (approval no.: 169-210617) of the hospital where the study was conducted. This study was conducted in accordance with the Declaration of Helsinki and ethical guidelines for medical and health research involving human subjects.

Informed Consent Statement: Patient consent was waived due to the constraints imposed by the retrospective study design, although the participants could withdraw from this study at any time by using an opt-out procedure.

Data Availability Statement: The data are not publicly available owing to opt out restrictions. Data sharing is not applicable.

Acknowledgments: We greatly appreciate the Nutritional Support Team at Kumamoto Rehabilitation Hospital who supported this study.

Conflicts of Interest: The authors declare no conflict of interest.

References

1. Strakowski, M.M.; Strakowski, J.A.; Mitchell, M.C. Malnutrition in rehabilitation. *Am. J. Phys. Med. Rehabil.* **2002**, *81*, 77–78. [CrossRef]
2. Wojzischke, J.; Van Wijngaarden, J.; Van den Berg, C.; Cetinyurek-Yavuz, A.; Diekmann, R.; Luiking, Y.; Bauer, J. Nutritional status and functionality in geriatric rehabilitation patients: A systematic review and meta-analysis. *Eur Geriatr. Med.* **2020**, *11*, 195–207. [CrossRef] [PubMed]
3. Yoshimura, Y.; Wakabayashi, H.; Bise, T.; Tanoue, M. Prevalence of sarcopenia and its association with activities of daily living and dysphagia in convalescent rehabilitation ward inpatients. *Clin. Nutr.* **2018**, *37 (Pt A)*, 2022–2028. [CrossRef]
4. Shiraishi, A.; Yoshimura, Y.; Wakabayashi, H.; Tsuji, Y. Prevalence of stroke-related sarcopenia and its association with poor oral status in post-acute stroke patients: Implications for oral sarcopenia. *Clin. Nutr.* **2018**, *37*, 204–207. [CrossRef]
5. Volkert, D.; Beck, A.M.; Cederholm, T.; Cruz-Jentoft, A.; Goisser, S.; Hooper, L.; Kiesswetter, E.; Maggio, M.; Raynaud-Simon, A.; Sieber, C.C.; et al. ESPEN guideline on clinical nutrition and hydration in geriatrics. *Clin. Nutr.* **2019**, *38*, 10–47. [CrossRef]
6. Yoshimura, Y.; Wakabayashi, H.; Bise, T.; Nagano, F.; Shimazu, S.; Shiraishi, A.; Yamaga, M.; Koga, H. Sarcopenia is associated with worse recovery of physical function and dysphagia and a lower rate of home discharge in Japanese hospitalized adults undergoing convalescent rehabilitation. *Nutrition* **2019**, *61*, 111–118. [CrossRef] [PubMed]
7. Yoshimura, Y.; Wakabayashi, H.; Nagano, F.; Bise, T.; Shimazu, S.; Kudo, M.; Shiraishi, A. Sarcopenic Obesity Is Associated With Activities of Daily Living and Home Discharge in Post-Acute Rehabilitation. *J. Am. Med. Dir. Assoc.* **2020**, *21*, 1475–1480. [CrossRef]
8. Nakahara, S.; Takasaki, M.; Abe, S.; Kakitani, C.; Nishioka, S.; Wakabayashi, H.; Maeda, K. Aggressive nutrition therapy in malnutrition and sarcopenia. *Nutrition* **2020**, *84*, 111109. [CrossRef] [PubMed]
9. Nishioka, S.; Aragane, H.; Suzuki, N.; Yoshimura, Y.; Fujiwara, D.; Mori, T.; Kanehisa, Y.; Iida, Y.; Higashi, K.; Yoshimura-Yokoi, Y.; et al. Clinical practice guidelines for rehabilitation nutrition in cerebrovascular disease, hip fracture, cancer, and acute illness: 2020 update. *Clin. Nutr. ESPEN* **2021**. [CrossRef] [PubMed]
10. Van Wijngaarden, J.P.; Wojzischke, J.; Van den Berg, C.; Cetinyurek-Yavuz, A.; Diekmann, R.; Luiking, Y.C.; Bauer, J.M. Effects of Nutritional Interventions on Nutritional and Functional Outcomes in Geriatric Rehabilitation Patients: A Systematic Review and Meta-Analysis. *J. Am. Med. Dir. Assoc* **2020**, *21*, 1207–1215.e9. [CrossRef] [PubMed]
11. Kokura, Y.; Nishioka, S.; Okamoto, T.; Takayama, M.; Miyai, I. Weight gain is associated with improvement in activities of daily living in underweight rehabilitation inpatients: A nationwide survey. *Eur. J. Clin. Nutr.* **2019**, *73*, 1601–1604. [CrossRef]
12. Nagano, F.; Yoshimura, Y.; Bise, T.; Shimazu, S.; Shiraishi, A. Muscle mass gain is positively associated with functional recovery in patients with sarcopenia after stroke. *J. Stroke Cerebrovasc. Dis.* **2020**, *29*, 105017. [CrossRef]
13. Walker, J.; Roberts, S.L.; Halmi, K.A.; Goldberg, S.C. Caloric requirements for weight gain in anorexia nervosa. *Am. J. Clin. Nutr.* **1979**, *32*, 1396–1400. [CrossRef]
14. Hall, K.D. What is the required energy deficit per unit weight loss? *Int. J. Obes.* **2008**, *32*, 573–576. [CrossRef] [PubMed]
15. Yoshimura, Y.; Wakabayashi, H.; Nagano, F.; Bise, T.; Shimazu, S.; Shiraishi, A. Elevated Creatinine-Based Estimated Glomerular Filtration Rate is Associated with Increased Risk of Sarcopenia, Dysphagia, and Reduced Functional Recovery after Stroke. *J. Stroke Cerebrovasc. Dis.* **2021**, *30*, 105491. [CrossRef] [PubMed]
16. Cederholm, T.; Jensen, G.L.; Correia, M.I.T.D.; Gonzalez, M.C.; Fukushima, R.; Higashiguchi, T.; Baptista, G.; Barazzoni, R.; Blaauw, R.; Coats, A.J.S.; et al. GLIM criteria for the diagnosis of malnutrition—A consensus report from the global clinical nutrition community. *Clin. Nutr.* **2019**, *38*, 1–9. [CrossRef]
17. Yoshimura, Y.; Wakabayashi, H.; Momosaki, R.; Nagano, F.; Shimazu, S.; Shiraishi, A. Shorter Interval between Onset and Admission to Convalescent Rehabilitation Wards Is Associated with Improved Outcomes in Ischemic Stroke Patients. *Tohoku J. Exp. Med.* **2020**, *252*, 15–22. [CrossRef]
18. Rubenstein, L.Z.; Harker, J.O.; Salvà, A.; Guigoz, Y.; Vellas, B. Screening for undernutrition in geriatric practice: Developing the short-form mini-nutritional assessment (MNA-SF). *J. Gerontol. Ser. A Biol. Sci. Med. Sci.* **2001**, *56*, M366–M372. [CrossRef]
19. Kunieda, K.; Ohno, T.; Fujishima, I.; Hojo, K.; Morita, T. Reliability and validity of a tool to measure the severity of dysphagia: The Food Intake LEVEL Scale. *J. Pain Symptom Manag.* **2013**, *46*, 201–206. [CrossRef]
20. Charlson, M.E.; Pompei, P.; Ales, K.L.; MacKenzie, C.R. A new method of classifying prognostic comorbidity in longitudinal studies: Development and validation. *J. Chronic Dis.* **1987**, *40*, 373–383. [CrossRef]

21. Banks, J.L.; Marotta, C.A. Outcomes validity and reliability of the modified Rankin scale: Implications for stroke clinical trials: A literature review and synthesis. *Stroke* **2007**, *38*, 1091–1096. [CrossRef]
22. Huang, C.Y.; Lin, G.H.; Huang, Y.J.; Song, C.-Y.; Lee, Y.-C.; How, M.-J.; Chen, Y.-M.; Hsueh, I.-P.; Chen, M.-H.; Hsieh, C.-L. Improving the utility of the Brunnstrom recovery stages in patients with stroke: Validation and quantification. *Medicine* **2016**, *95*, e4508. [CrossRef]
23. Ottenbacher, K.J.; Hsu, Y.; Granger, C.V.; Fiedler, R.C. The reliability of the functional independence measure: A quantitative review. *Arch. Phys. Med. Rehabil.* **1996**, *77*, 1226–1232. [CrossRef]
24. Yoshimura, Y.; Wakabayashi, H.; Shiraishi, A.; Nagano, F.; Bise, T.; Shimazu, S. Hemoglobin Improvement is Positively Associated with Functional Outcomes in Stroke Patients with Anemia. *J. Stroke Cerebrovasc. Dis.* **2021**, *30*, 105453. [CrossRef]
25. Chen, L.K.; Woo, J.; Assantachai, P.; Auyeung, T.-W.; Chou, M.-Y.; Iijima, K.; Jang, H.C.; Kang, L.; Kim, M.; Kim, S.; et al. Asian Working Group for Sarcopenia: 2019 Consensus Update on Sarcopenia Diagnosis and Treatment. *J. Am. Med. Dir. Assoc.* **2020**, *21*, 300–307.e2. [CrossRef] [PubMed]
26. Sawaya, A.L.; Tucker, K.; Tsay, R.; Willett, W.; Saltzman, E.; Dallal, G.E.; Roberts, S.B. Evaluation of four methods for determining energy intake in young and older women: Comparison with doubly labeled water measurements of total energy expenditure. *Am. J. Clin. Nutr.* **1996**, *63*, 491–499. [CrossRef] [PubMed]
27. Burgos, R.; Bretón, I.; Cereda, E.; Desport, J.C.; Dziewas, R.; Genton, L.; Gomes, F.; Jésus, P.; Leischker, A.; Muscaritoli, M.; et al. ESPEN guideline clinical nutrition in neurology. *Clin. Nutr.* **2018**, *37*, 354–396. [CrossRef]
28. Yoshimura, Y.; Wakabayashi, H.; Nagano, F.; Bise, T.; Shimazu, S.; Shiraishi, A. Chair-stand exercise improves post-stroke dysphagia. *Geriatr. Gerontol. Int.* **2020**, *20*, 885–891. [CrossRef] [PubMed]
29. Kido, Y.; Yoshimura, Y.; Wakabayashi, H.; Momosaki, R.; Nagano, F.; Bise, T.; Shimazu, S.; Shiraishi, A. Sarcopenia is associated with incontinence and recovery of independence in urination and defecation in post-acute rehabilitation patients. *Nutrition* **2021**, *91–92*, 111397. [CrossRef]
30. Shimazu, S.; Yoshimura, Y.; Kudo, M.; Nagano, F.; Bise, T.; Shiraishi, A.; Sunahara, T. Frequent and personalized nutritional support leads to improved nutritional status, activities of daily living, and dysphagia after stroke. *Nutrition* **2021**, *83*, 111091. [CrossRef] [PubMed]
31. Shiraishi, A.; Wakabayashi, H.; Yoshimura, Y. Oral Management in Rehabilitation Medicine: Oral Frailty, Oral Sarcopenia, and Hospital-Associated Oral Problems. *J. Nutr. Health Aging* **2020**, *24*, 1094–1099. [CrossRef]
32. Shiraishi, A.; Yoshimura, Y.; Wakabayashi, H.; Tsuji, Y.; Yamaga, M.; Koga, H. Hospital dental hygienist intervention improves activities of daily living, home discharge and mortality in post-acute rehabilitation. *Geriatr. Gerontol. Int.* **2019**, *19*, 189–196. [CrossRef] [PubMed]
33. Kose, E.; Wakabayashi, H. Rehabilitation pharmacotherapy: A scoping review. *Geriatr Gerontol Int.* **2020**, *20*, 655–663. [CrossRef] [PubMed]
34. Shiraisi, A.; Yoshimura, Y.; Wakabayashi, H.; Nagano, F.; Bise, T.; Shimazu, S. Improvement in Oral Health Enhances the Recovery of Activities of Daily Living and Dysphagia after Stroke. *J. Stroke Cerebrovasc. Dis.* **2021**, *30*, 105961. [CrossRef] [PubMed]
35. Lieber, A.C.; Hong, E.; Putrino, D.; Nistal, D.A.; Pan, J.S.; Kellner, C.P. Nutrition, energy expenditure, dysphagia, and self-efficacy in stroke rehabilitation: A review of the literature. *Brain Sci.* **2018**, *8*, 218. [CrossRef]
36. Scott, D.; Blizzard, L.; Fell, J.; Jones, G. Statin therapy, muscle function and falls risk in community-dwelling older adults. *QJM Int. J. Med.* **2009**, *102*, 625–633. [CrossRef] [PubMed]
37. Lin, M.-H.; Chiu, S.-Y.; Chang, P.-H.; Lai, Y.-L.; Chen, P.-C.; Ho, W.-C. Hyperlipidemia and Statins Use for the Risk of New Diagnosed Sarcopenia in Patients with Chronic Kidney: A Population-Based Study. *Int. J. Environ. Res. Public Health* **2020**, *17*, 1494. [CrossRef] [PubMed]
38. Hébuterne, X.; Bermon, S.; Schneider, S.M. Ageing and muscle: The effects of malnutrition, re-nutrition, and physical exercise. *Curr. Opin. Clin. Nutr. Metab. Care* **2001**, *4*, 295–300. [CrossRef]
39. Nagano, A.; Nishioka, S.; Wakabayashi, H. Rehabilitation Nutrition for Iatrogenic Sarcopenia and Sarcopenic Dysphagia. *J. Nutr. Health Aging* **2019**, *23*, 256–265. [CrossRef]

Article

Physical Activity and Diet Quality Modify the Association between Comorbidity and Disability among Stroke Patients

Lien T. K. Nguyen [1,2,3], Binh N. Do [4,5], Dinh N. Vu [6,7], Khue M. Pham [8,9], Manh-Tan Vu [10,11], Hoang C. Nguyen [12,13], Tuan V. Tran [14,15], Hoang P. Le [16], Thao T. P. Nguyen [17,18], Quan M. Nguyen [19], Cuong Q. Tran [20,21], Kien T. Nguyen [22], Shwu-Huey Yang [23,24,25], Jane C.-J. Chao [23,24,26] and Tuyen Van Duong [23,*]

[1] Rehabilitation Department, Hanoi Medical University, Hanoi 115-20, Vietnam; lienrehab@hmu.edu.vn
[2] Rehabilitation Center, Bach Mai Hospital, Hanoi 115-19, Vietnam
[3] Rehabilitation Department, Viet Duc University Hospital, Hanoi 110-17, Vietnam
[4] Department of Infectious Diseases, Vietnam Military Medical University, Hanoi 121-08, Vietnam; nhubinh.do@vmmu.edu.vn
[5] Division of Military Science, Military Hospital 103, Hanoi 121-08, Vietnam
[6] Director Office, Military Hospital 103, Hanoi 121-08, Vietnam; vunhatdinh@vmmu.edu.vn
[7] Department of Trauma and Orthopedic Surgery, Vietnam Military Medical University, Hanoi 121-08, Vietnam
[8] Faculty of Public Health, Hai Phong University of Medicine and Pharmacy, Hai Phong 042-12, Vietnam; pmkhue@hpmu.edu.vn
[9] President Office, Hai Phong University of Medicine and Pharmacy, Hai Phong 042-12, Vietnam
[10] Department of Internal Medicine, Haiphong University of Medicine and Pharmacy, Hai Phong 042-12, Vietnam; vmtan@hpmu.edu.vn
[11] Cardiovascular Department, Viet Tiep Friendship Hospital, Hai Phong 047-08, Vietnam
[12] Director Office, Thai Nguyen National Hospital, Thai Nguyen City 241-24, Vietnam; nguyenconghoang@tnmc.edu.vn
[13] President Office, Thai Nguyen University of Medicine and Pharmacy, Thai Nguyen City 241-17, Vietnam
[14] Department of Neurology, Thai Nguyen University of Medicine and Pharmacy, Thai Nguyen City 241-17, Vietnam; tranvantuanyktn@gmail.com
[15] Department of Clinical Pharmacy, Thai Nguyen University of Medicine and Pharmacy, Thai Nguyen City 241-17, Vietnam
[16] Department of Internal Medicine, University of Medicine and Pharmacy, Hue University, Thua Thien Hue 491-20, Vietnam; lphoang@huemed-univ.edu.vn
[17] Health Management Training Institute, University of Medicine and Pharmacy, Hue University, Thua Thien Hue 491-20, Vietnam; nguyenthiphuongthao@hueuni.edu.vn
[18] Department of Health Economics, Corvinus University of Budapest, 1093 Budapest, Hungary
[19] Director Office, Thu Duc City Hospital, Ho Chi Minh City 713-11, Vietnam; quan_minhnguyen@yahoo.com
[20] Director Office, Thu Duc City Health Center, Ho Chi Minh City 713-10, Vietnam; quoccuong.mph@gmail.com
[21] Faculty of Health, Mekong University, Vinh Long 852-16, Vietnam
[22] Department of Health Promotion, Faculty of Social and Behavioral Sciences, Hanoi University of Public Health, Hanoi 119-10, Vietnam; ntk1@huph.edu.vn
[23] School of Nutrition and Health Sciences, Taipei Medical University, Taipei 110-31, Taiwan; sherry@tmu.edu.tw (S.-H.Y.); chenjui@tmu.edu.tw (J.C.-J.C.)
[24] Nutrition Research Center, Taipei Medical University Hospital, Taipei 110-31, Taiwan
[25] Research Center of Geriatric Nutrition, Taipei Medical University, Taipei 110-31, Taiwan
[26] Master Program in Global Health and Development, College of Public Health, Taipei Medical University, Taipei 110-31, Taiwan
* Correspondence: tvduong@tmu.edu.tw Tel.: +88-62-2736-1661 (ext. 6545)

Citation: Nguyen, L.T.K.; Do, B.N.; Vu, D.N.; Pham, K.M.; Vu, M.-T.; Nguyen, H.C.; Tran, T.V.; Le, H.P.; Nguyen, T.T.P.; Nguyen, Q.M.; et al. Physical Activity and Diet Quality Modify the Association between Comorbidity and Disability among Stroke Patients. *Nutrients* **2021**, *13*, 1641. https://doi.org/10.3390/nu13051641

Academic Editor: Yoshihiro Yoshimura

Received: 15 April 2021
Accepted: 12 May 2021
Published: 13 May 2021

Publisher's Note: MDPI stays neutral with regard to jurisdictional claims in published maps and institutional affiliations.

Abstract: Background: Comorbidity is common and causes poor stroke outcomes. We aimed to examine the modifying impacts of physical activity (PA) and diet quality on the association between comorbidity and disability in stroke patients. Methods: A cross-sectional study was conducted on 951 stable stroke patients in Vietnam from December 2019 to December 2020. The survey questionnaires were administered to assess patients' characteristics, clinical parameters (e.g., Charlson Comorbidity Index items), health-related behaviors (e.g., PA using the International Physical Activity Questionnaire- short version), health literacy, diet quality (using the Dietary Approaches to Stop Hypertension Quality (DASH-Q) questionnaire), and disability (using the World Health Organization Disability Assessment Schedule II (WHODAS II)). Linear regression models were used to analyze the associations and interactions. Results: The proportion of comorbidity was 49.9% (475/951). The scores of DASH-Q and WHODAS II were 29.2 ± 11.8, 32.3 ± 13.5, respectively. Patients with comorbidity had a higher score of disability (regression coefficient, B, 8.24; 95% confidence interval, 95%CI, 6.66, 9.83; $p < 0.001$) as compared with those without comorbidity. Patients with comorbidity and higher tertiles of PA (B, −4.65 to −5.48; $p < 0.05$), and a higher DASH-Q score (B, −0.32; $p < 0.001$) had a lower disability score, as compared with those without comorbidity and the lowest tertile of PA, and the lowest score of DASH-Q, respectively. Conclusions: Physical activity and diet quality significantly modified the negative impact of comorbidity on disability in stroke patients. Strategic approaches are required to promote physical activity and healthy diet which further improve stroke rehabilitation outcomes.

Keywords: stroke patient; Charlson Comorbidity Index; World Health Organization Disability Assessment Schedule II; international physical activity questionnaire; Dietary Approaches to Stop Hypertension Quality; health literacy; International Classification of Diseases; health-related behaviors; Vietnam

1. Introduction

Stroke is a major cause of disability and mortality across the globe [1]. In 2013, there were 113 million disability-adjusted life years, and 6.5 million deaths [2]. In Vietnam, the intracerebral hemorrhage stroke prevalence was as high as that in high-income countries [3]. Stroke and its consequences impose a heavy burden on individuals, the healthcare system, and society in Vietnam and the world [4–6].

The risk factors of stroke could be attributed to about 90% modifiable risks, including hypertension, obesity, hyperglycemia, hyperlipidemia, and renal dysfunction [6,7]. Comorbid conditions (or modifiable risks) are common in stroke patients [8,9]. They are predictors of hospital stay, costs, and mortality [10], increase the disability levels [11], and worsen functional outcomes after stroke [12].

Multidisciplinary and multilevel prevention strategies were suggested to prevent stroke. Among those, adequate nutrition, salt reduction, and other dietary interventions are effective strategies for primordial and primary prevention [13]. The longitudinal effect of diet quality (assessed by the Dietary Approaches to Stop Hypertension, or DASH diet) on cardiovascular diseases (CVD), including stroke, was summarized in a meta-analysis of prospective studies [14]. The relationship between diet quality and risk of stroke was also found in a previous large cohort study in European countries [15,16], Taiwan [17], and Hong Kong [18]. The key health behaviors (e.g., diet, physical activity) significantly contribute to cardiovascular conditions (including stroke and other heart and circulatory diseases) [7,19].

The roles of health-related behavioral factors (e.g., physical activity and dietary intake) on stroke prevention were investigated [6,7,19]. However, the modification effects of these factors remain to be explored. It is necessary to explore the potential impacts of health-related behaviors which may modify the negative effects of comorbidity on physical function after stroke. Therefore, we aimed to investigate the modifying impacts of physical activity and diet quality on the relationship between comorbidity and disability among stroke patients.

2. Materials and Methods

2.1. Study Design and Settings

A cross-sectional study design was used to survey stroke patients between December 2019 and December 2020 in four hospitals in the northern area, one hospital in the central area, and one hospital in the southern area of Vietnam.

2.2. Sampling and Sample Size

We used the consecutively convenient sampling technique to recruit patients from cardiovascular, neurology, and rehabilitation departments of selected hospitals. Data of 951 patients were collected from Bach Mai Hospital (11 from the cardiovascular department, 131 from the neurology department, 27 from the rehabilitation center), Military Hospital 103 (293 from the stroke department), Viet Tiep Friendship Hospital (197 from the neurology department), Thai Nguyen National Hospital (197 from the neurology department), Hue University Hospital (45 from the cardiovascular department), and Thu Duc District Hospital (50 from the neurology department).

Patients recruited were those aged $\geq$ 18 years, in a stable stroke condition diagnosed by a neurologist (e.g., a Mini-Mental State Examination score of $\geq$22), with the ability to respond to questions. Patients excluded were those with aphasia or visual impairment, or those with diseases that affect cognition (e.g., dementia). Stroke or cerebrovascular disease was defined by the 10th revision of the International Classification of Diseases (ICD-10) codes I60–69; including (I60) Subarachnoid hemorrhage; (I61) Intracerebral hemorrhage; (I62) Other non-traumatic intracranial hemorrhages; (I63) Cerebral infarction; (I64) Stroke, not specified as hemorrhage or infarction; (I65) Occlusion and stenosis of pre-cerebral arteries, not resulting in cerebral infarction; (I66) Occlusion and stenosis of cerebral arteries, not resulting in cerebral infarction; (I67) Other cerebrovascular diseases; (I68) Cerebrovascular disorders in diseases classified elsewhere; and (I69) Sequelae of cerebrovascular disease.

2.3. Measurements

2.3.1. Patients' Characteristics

Participants were asked about their age (years), gender (women vs. men), education attainment (illiterate/elementary school level, junior high school level, senior high school level, college/university level and above), marital status (married vs. single or separated/divorced/widowed), occupation (working vs. retired or infirmity), ability to pay for medication (very or fairly difficult vs. very or fairly easy), and their social status level (low vs. middle or high).

2.3.2. Health-Related Behaviors

Patients were asked about their current behaviors, regarding the status of smoking (never vs. ever smoked), drinking alcohol (no vs. yes).

The International Physical Activity Questionnaire short version (IPAQ-SF) was used for assessing physical activity (PA) level [20]. Patients reported their time spent on different PA types over the last seven days. The IPAQ was validated and used in the Vietnamese population [21]. The overall PA score was calculated by multiplying minutes spent on activities at different levels including vigorous, moderate, walking, and sitting by 8.0, 4.0, 3.3, or 1.0, respectively [20]. The metabolic equivalent task scored in minutes per week (MET-min/wk) was used as the measuring unit of PA [22].

2.3.3. Clinical Parameters

Comorbidity was assessed using the Charlson Comorbidity Index (CCI) items [23,24]. The item list consists of (1) myocardial infarction (history, not ECG changes only); (2) congestive heart failure; (3) peripheral disease (includes aortic aneurysm $\geq$ 6 cm); (4) cerebrovascular disease or stroke; (5) chronic pulmonary disease; (6) diabetes without end-organ damage (excludes diet-controlled alone); (7) depression; (8) diseases treated with anticoagulants; (9) dementia; (10) hemiplegia; (11) diabetes with end-organ damages (retinopathy, neuropathy, nephropathy, or brittle diabetes); (12) moderate or severe renal diseases; (13) tumor without metastasis (exclude if > 5 years from diagnosis), leukemia (acute or chronic) and lymphoma; (14) moderate or severe liver diseases; (15) metastatic solid tumor; and (16) HIV/AIDS. Patients with dementia were excluded. Cerebrovascular disease, although reported in the list, was excluded when calculating the number of chronic conditions. The comorbidity was classified into two groups (none vs. one or more) to facilitate the analysis.

The body mass index (BMI), a measure of body fat, was calculated as weight (kg)/[height (m)]^2. The stroke occurrence (first ever vs. recurrent) was also assessed.

2.3.4. Health Literacy

Health literacy (HL) was assessed using the short-form questionnaire with 12 items (HLS-SF12). It was validated and used in Asian countries [25,26], including Vietnam [27–30]. Patients were asked to rate their perceived difficulty of each item based on 4-point Likert scales from 1 = "very difficult" to 4 = "very easy". The overall score was standardized to an index ranging from 0 to 50 using Formula (1), with a higher score presenting better HL [31]:

$$\text{Index} = (\text{Mean} - 1) \times (50/3) \tag{1}$$

where Index is a specific index score calculated, Mean is the mean of 12 items, 1 is the minimal possible value of the mean (leading to a minimum index score of 0), 3 is the range of the mean, and 50 is the chosen maximum HL index score.

2.3.5. Diet Quality

A brief self-reported measure of diet quality questionnaire, or the Dietary Approaches to Stop Hypertension Quality (DASH-Q) questionnaire, was used for assessing diet quality [32]. The DASH-Q consists of 11 items and asks how many days, over the past 7 days, did patients eat the food items. The response options are from 0 to 7. Two researchers translated the questionnaire into the Vietnamese language. An expert panel (five medical doctors, five public health and nutrition professionals) then validated the content and suggested keeping the original response options and scoring. In Table S1, item "drink milk (in a glass, with cereal, or in coffee, tea, or cocoa)" was removed with a factor loading of −0.45 on component 2, and 0.41 on component 3. The items were loaded on three components, which explained 62.76% of the variance. The DASH-Q with 10 items illustrated adequate convergent validity (item-subscale correlation ranges of 0.70–0.80 on factor 1 and component 3, 0.74–0.89 on component 2,), satisfactory reliability (Cronbach's alpha of 0.74), and no floor or ceiling effects (Table S1). A sum score of DASH-Q was recommended for use in research and clinical practices. The overall DASH-Q scores range from 0 to 70, with higher scores presenting better diet quality. Participants with high diet quality were considered adherent to DASH nutritional recommendations [32].

2.3.6. Disability

The 12-item World Health Organization Disability Assessment Schedule II (named as WHODAS II) was used for assessing the disability level in different cultures and settings [33,34]. Patients were asked to rate on 5-point scales the extent of difficulty of doing the activities in the past 30 days from (1) none to (2) mild, (3) moderate, (4) severe, (5) extreme difficulty or cannot do [35]. The overall WHODAS II score was a sum score of all items, with a higher score reflecting a higher disability level.

2.4. Data Collection Procedure

Research assistants (e.g., doctors, nurses, and medical students) firstly received a four-hour training session about data collection, conducted by researchers from each hospital. Doctors in charge selected qualified patients for the study. Research assistants then approached and asked for patients' voluntary participation. The informed consent form was signed by patients before administering the survey. The face-to-face interviews were conducted at the bedside. Adequate time was given to patients to complete the survey. It took about 30 min to complete a survey for one patient. After the interview, research assistants reviewed the medical records for clinical parameters.

We postponed the data collection during the COVID-19-induced nationwide lockdown in Vietnam from 1st to 22nd of April 2020 [36,37]. During the pandemic, research assistants also received infection control training from each hospital, in terms of mask-wearing, handwashing, and physical distancing which followed the guidelines of the Centers for

Disease Control and Prevention (CDC) [38], World Health Organization (WHO) [39], and Ministry of Health in Vietnam [40].

2.5. Ethical Consideration

The study was reviewed and approved by the Institutional Ethical Review Committee of Hanoi University of Public Health (IRB No. 498/2019/YTCC-HD3 and No. 312/2020/YTCC-HD3). All subjects gave their informed consent for inclusion before they participated in the study.

2.6. Statistical Analysis

Firstly, the studied variables' distributions were explored using descriptive analysis. A one-way ANOVA test was used to compare the distribution of WHODAS II in different categories of independent variables. Secondly, the associations of comorbidity, physical activity, and diet quality (DASH-Q) with disability (WHODAS II) were analyzed using bivariate and multivariate linear regression models. We adjusted for age, gender, marital status, education, occupation, smoking status, and health literacy in multivariate linear regression models as these variables showed associations with WHODAS II in the bivariate linear regression models (Table S2). Finally, the interaction analysis was performed to examine the potential modification impacts of physical activity and diet quality on the association between comorbidity and disability. To visualize the results of the interaction models, we conducted a simple slope analysis using PROCESS Macro of SPSS for moderation analysis [41]. The slope plots were drawn using the evaluated values of WHODAS II for two categories of comorbidity (non-CCI vs. CCI) by three values of physical activity (mid-tertile 1, mid-tertile 2, and mid-tertile 3 of MET-min/wk), and diet quality ($Z = -1$, one standard deviation below the mean; $Z = 0$, the mean; $Z = +1$, 1 standard deviation above the mean of DASH-Q). Data were analyzed using IBM SPSS Version 20.0 (IBM Corp, Armonk, NY, USA) [42]. The significance level was set at a p-value < 0.05.

3. Results

3.1. Patients' Characteristics

In the total sample, 70% of stroke patients were 60 years old or above, 59.2% were men. The proportions of patients with comorbid and first-ever stroke were 49.9% and 82.5%, respectively. The scores of DASH-Q, HL, and WHODAS II were 29.2 ± 11.8, 23.4 ± 10.0, 32.3 ± 13.5, respectively. The score of WHODAS II was significantly varied in different categories of age, gender, marital status, education, occupation, comorbidity, smoking, and physical activity (Table 1).

Table 1. Characteristics and disability in stroke patients ($n = 951$).

Variables	Total	WHODAS II	p *
	n (%)	(Mean $\pm$ SD)	
Age, years			<0.001
19–59	285 (30.0)	30.4 ± 13.1	
60–69	286 (30.1)	31.2 ± 13.5	
70–79	222 (23.3)	33.3 ± 13.4	
80–99	158 (16.6)	36.5 ± 13.3	
Gender			0.035
Women	388 (40.8)	33.5 ± 13.3	
Men	563 (59.2)	31.6 ± 13.5	

Table 1. *Cont.*

Variables	Total	WHODAS II	p *
	n (%)	(Mean ± SD)	
Marital status			0.040
Married	837 (88.0)	32.0 ± 13.6	
Single or Widowed/Divorced/Separated	114 (12.0)	34.8 ± 12.4	
Education attainment			0.002
Illiterate or elementary	215 (22.6)	34.4 ± 12.3	
Junior high	257 (27.1)	33.6 ± 14.3	
Senior high	251 (26.4)	31.4 ± 14.5	
College/university or higher	227 (23.9)	30.1 ± 12.0	
Occupation			<0.001
Working	518 (54.5)	29.6 ± 12.9	
Retired or infirmity	433 (45.5)	35.6 ± 13.4	
Ability to pay for medication			0.528
Very or fairly difficult	423 (44.5)	32.7 ± 13.7	
Very or fairly easy	528 (55.5)	32.1 ± 13.3	
Social status			0.344
Low	111 (11.7)	33.5 ± 12.3	
Middle or high	840 (88.3)	32.2 ± 13.6	
BMI, kg/m^2			0.203
Underweight (<18.5)	90 (9.5)	33.5 ± 14.6	
Normal weight (18.5 ≤ BMI < 24.0)	794 (83.7)	32.4 ± 13.4	
Overweight/obese (BMI ≥ 25.0)	65 (6.8)	29.7 ± 12.8	
CCI			<0.001
None	476 (50.1)	28.1 ± 11.8	
One or more	475 (49.9)	36.6 ± 13.8	
Stroke occurrence			0.261
First ever	785 (82.5)	32.1 ± 13.2	
Recurrent	166 (17.5)	33.4 ± 14.8	
Smoking			0.028
Never smoked	544 (57.2)	31.5 ± 13.5	
Ever smoked	407 (42.8)	33.5 ± 13.4	
Drinking alcohol			0.071
No	661 (69.5)	32.9 ± 13.7	
Yes	290 (30.5)	31.2 ± 13.0	
Physical activity, MET-min/wk			<0.001
Tertile 1 (MET ≤ 597)	324 (34.1)	38.4 ± 13.3	
Tertile 2 (597 < MET ≤ 3726)	312 (32.8)	31.0 ± 13.2	
Tertile 3 (MET > 3726)	315 (33.1)	27.5 ± 11.6	
DASH-Q, mean ± SD	29.2 ± 11.8		
HL index, mean ± SD	23.4 ± 10.0		
WHODAS II, mean ± SD	32.3 ± 13.5		

Abbreviation: SD, standard deviation; WHODAS II, World Health Organization Disability Assessment Schedule II; BMI, body mass index; CCI, Charlson Comorbidity Index; MET-min/wk, metabolic equivalent task scored in minutes per week; DASH-Q, Dietary Approaches to Stop Hypertension Quality; HL, health literacy.* Results of one-way ANOVA test.

3.2. Associations of Comorbidity, Physical Activity, Diet Quality with Disability

The results of the multivariate analysis (after adjusting for age, gender, marital status, education attainment, occupation, smoking status, and health literacy) illustrate that patients with comorbidity had a higher disability score (regression coefficient, B, 8.24; 95% confidence interval, 95%CI, 6.66, 9.83; $p < 0.001$) as compared with those without comorbidity. In comparison with patients' exercise level in the first tertile, those in the second tertile (B, -6.49; 95%CI, -8.51, -4.47; $p < 0.001$), or third tertile (B -9.00; 95%CI, -11.06, -6.94; $p < 0.001$) had a lower score of disability. Patients with a one-point increment in DASH-Q had a 0.20-point reduction in disability (B, -0.20; 95%CI, -0.27, -0.13; $p < 0.001$; Table 2).

Table 2. Associations of comorbidity, physical activity, and diet quality with disability among stroke patients ($n = 951$).

Variables	WHODAS II		WHODAS II	
	B (95%CI) *	*p*	B (95%CI) **	*p*
CCI				
None	Reference		Reference	
One or more	8.51 (6.88, 10.14)	<0.001	8.24 (6.66, 9.83)	<0.001
Physical activity, MET-min/wk				
Tertile 1	Reference			
Tertile 2	-7.40 (-9.37, -5.42)	<0.001	-6.49 (-8.51, -4.47)	<0.001
Tertile 3	-10.82 (-12.80, -8.85)	<0.001	-9.00 (-11.06, -6.94)	<0.001
DASH-Q, 1-point increment	-0.27 (-0.34, -0.20)	<0.001	-0.20 (-0.27, -0.13)	<0.001

Abbreviation: WHODAS II, World Health Organization Disability Assessment Schedule II; CCI, Charlson Comorbidity Index; MET-min/wk, metabolic equivalent task scored in minutes per week; DASH-Q, Dietary Approaches to Stop Hypertension Quality.* Results of bivariate linear regression analysis.** Results of multivariate linear regression analysis after adjustment for age, gender, marital status, education attainment, occupation, smoking status, and health literacy.

3.3. Modification Impacts of Physical Activity, Diet Quality

In the multivariate analysis, in comparison with patients with no comorbidity and in the lowest tertile of physical activity (PA), those with a comorbidity and in the lowest tertile of PA had scores of disability 10.67 points higher (B, 10.67; 95%CI, 7.96, 13.37; $p < 0.001$), and those in the second tertile of PA (B, -4.65; 95%CI, -8.44, -0.85; $p < 0.016$), and third tertile of PA (B, -5.48; 95%CI, -9.27, -1.70; $p < 0.005$) had scores of disability 4.65 and 5.48 points lower, respectively (Table 3). The model results are illustrated in Figure 1.

Table 3. Interactions of comorbidity with physical activity and diet quality on disability among stroke patients ($n = 951$).

Interactions	WHODAS II		WHODAS II	
	B (95%CI) *	*p*	B (95%CI) **	*p*
CCI and MET				
Non-CCI × MET (tertile 1)	Reference		Reference	
CCI × MET (tertile 1)	10.70 (7.93, 13.47)	<0.001	10.67 (7.96, 13.37)	<0.001
Non-CCI × MET (tertile 2)	-2.74 (-5.59, 0.11)	0.059	-2.50 (-5.31, 0.32)	0.082
Non-CCI × MET (tertile 3)	-6.23 (-9.08, -3.39)	<0.001	-4.56 (-7.40, -1.72)	0.002
CCI × MET (tertile 2)	-5.47 (-9.37, -1.58)	0.006	-4.65 (-8.44, -0.85)	0.016
CCI × MET (tertile 3)	-5.39 (-9.27, -1.50)	0.007	-5.48 (-9.27, -1.70)	0.005
CCI and DASH-Q				
Non-CCI × DASH-Q (lowest score)				
CCI × DASH-Q (lowest score)	18.62 (14.36, 22.88)	<0.001	17.12 (12.98, 21.45)	<0.001
Non-CCI × DASH-Q (1-point increment)	-0.03 (-0.13, 0.07)	0.523	-0.01 (-0.09, 0.09)	0.990
CCI × DASH-Q (1-point increment)	-0.37 (-0.51, -0.24)	<0.001	-0.32 (-0.45, -0.19)	<0.001

Abbreviations: WHODAS II, World Health Organization Disability Assessment Schedule II; CCI, Charlson Comorbidity Index; MET, metabolic equivalent task scored in minutes per week; DASH-Q, Dietary Approaches to Stop Hypertension Quality.* Results of bivariate linear regression analysis.** Results of multivariate linear regression analysis adjusted for age, gender, marital status, education, occupation, smoking status, and health literacy.

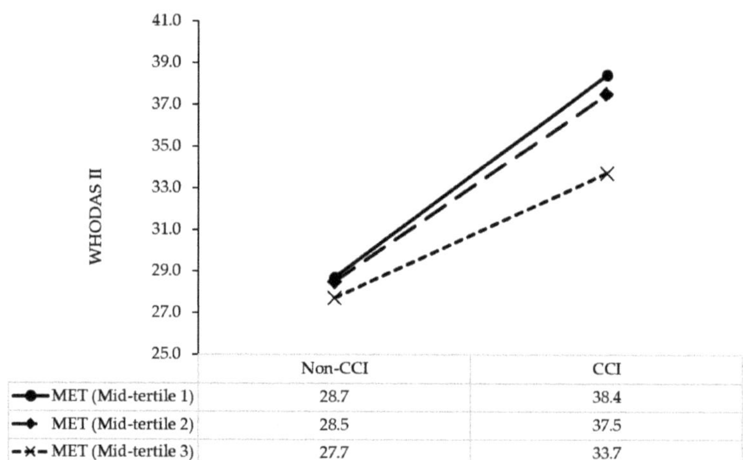

	Non-CCI	CCI
—●— MET (Mid-tertile 1)	28.7	38.4
—◆ MET (Mid-tertile 2)	28.5	37.5
–✕– MET (Mid-tertile 3)	27.7	33.7

Figure 1. Simple slope plot for the interaction between comorbidity and physical activity on disability among stroke patients ($n = 951$). CCI, Charlson Comorbidity Index; WHODAS II, World Health Organization Disability Assessment Schedule II; MET, metabolic equivalent task scored in minutes per week.

Similarly, in comparison with patients with no comorbidity and the lowest score of DASH-Q, those with a comorbidity and the lowest score of DASH-Q had scores of disability 17.12 points higher (B, 17.12; 95%CI, 12.98, 21.45; $p < 0.001$), and those with a one-point increment in DASH-Q had scores of disability 0.32 points lower (B, -0.32; 95%CI, -0.45, -0.19; $p < 0.001$; Table 3). The model results are illustrated in Figure 2.

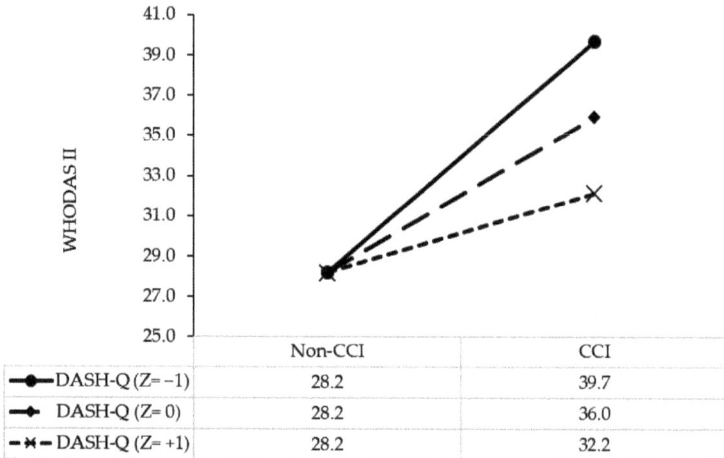

	Non-CCI	CCI
—●— DASH-Q ($Z = -1$)	28.2	39.7
—◆ DASH-Q ($Z = 0$)	28.2	36.0
–✕– DASH-Q ($Z = +1$)	28.2	32.2

Figure 2. Simple slope plot for the interaction between comorbidity and diet quality on disability among stroke patients ($n = 951$). Note: $Z = -1$, one standard deviation below the mean; $Z = 0$, the mean; $Z = +1$, 1 standard deviation above the mean. CCI, Charlson Comorbidity Index; WHODAS II, World Health Organization Disability Assessment Schedule II; DASH-Q, Dietary Approaches to Stop Hypertension Quality.

4. Discussion

In this study, comorbidity was found as one of the key predictors of disability in stroke patients. Comorbid medical conditions were strongly associated with poor outcomes and death [43–45], with higher disability levels [11], and worse functional outcomes after stroke [12]. It is necessary to evaluate the comorbid conditions in order to help develop appropriate plans for treatment and rehabilitation.

A higher physical activity score was associated with a lower disability score in the current study. Daily physical activity was found to be independently associated with a better physical component of quality of life in stroke patients, in a previous study [46]. Stroke patients spent more time on sedentary behaviors which further affect physical function [47–49] and recovery after stroke [50]. The prevalence of physical inactivity was increased during the COVID-19 pandemic which further created a huge burden of cardiovascular disease [51,52]. Therefore, physical activity should be promoted to potentially reduce physical limitation or disability in people living with stroke and improve the outcomes of rehabilitation therapy.

A healthy diet has been found to exert beneficial effects on cardiovascular disease (CVD) prevention [53–55]. The diet quality was associated with CVD-free life expectancy [56]. However, the certainty of the evidence was low [53,57]. In our current study, a higher score of diet quality was also associated with a lower disability score. However, a previous study showed no significant role of nutrition therapy in activities of daily living (ADL) in older stroke patients [58]. In a 5.3-year follow-up study conducted on older adults, a healthy diet (e.g., DASH diet) was associated with a lower likelihood of ADL disability and mobility disability [59]. In addition, adherence to a healthy diet was associated with a lower frailty index, reflecting an aspect of disability in older adults [60,61]. Moreover, in the older population, a healthy diet has also shown positive associations with muscle mass, muscle strength, and physical performance [62]. Therefore, it is suggested that a healthy diet may be an effective approach for the prevention of malnutrition and functional disability in older people, especially in those with stroke. A previous study showed that early nutritional intake after acute stroke admission had positive impacts on home discharge and ADL [63].

A healthy diet has been linked with lower concentrations of inflammatory parameters as risk factors for cardiovascular diseases [64]. During the COVID-19 pandemic, nutrition therapy has been found to be a potential protective approach to support the immune system, and reduce inflammation [65], while several nutrients may enhance the functional status [66]. Moreover, patients with better nutritional status had a lower thrombosis incidence which is recognized as a risk of cerebrovascular disease [67], e.g., stroke caused by COVID-19 infection [68]. Equally important to a healthy diet, physical activity has the potential to improve inflammatory status and mobility in chronic stroke patients [69]. Inflammation is common in several chronic diseases [70]. It shows an association with malnutrition, functional outcomes [71], and disability [72].

In the current study, physical activity and diet quality significantly modified the negative impacts of comorbidity on disability in stroke patients. A previous study has shown that being physically active and eating a healthy diet were associated with lower disability-adjusted life years (DALYs), and those who adhere to more healthy behaviors lived longer in good health [73]. During the COVID-19 pandemic, the adherence to a healthy diet slightly increased [74,75], although unhealthy food consumption and physical inactivity also increased [74,76]. Importantly, both a healthy diet and physical activity have shown benefits for first and recurrent stroke prevention [77]. Therefore, promoting a life-long healthy lifestyle is the most important way to primarily prevent CVD, as emphasized in the updated guidelines for primary CVD prevention [78].

Our study has some limitations. Firstly, the study sample was relatively small, which limits the analysis of the impact of the individual comorbid conditions on disability. Secondly, the duration of different comorbid diseases was not investigated in the current study, which affected the analysis of the association. Thirdly, the potential factors that may render food intake difficult (i.e., dysphagia after stroke, modification of diet texture) were not investigated in our study which might confound the findings. Finally, the generalizability and causality cannot be inferred from a cross-sectional design with consecutive convenient sampling. For example, it is also possible that disability affects food availability (e.g., no access to fresh food, no possibility to cook), which may explain the observed correlation. A longitudinal design is required to confirm the association. Despite the abovementioned limitations, the findings indicate a phenomenon and direction for future research and provide evidence for strategic interventions which may alleviate the negative impact of the comorbid condition on disability in stroke patients.

5. Conclusions

In stroke patients, comorbidity, physical activity, and diet quality were significantly associated with disability status. Importantly, the negative impact of comorbidity on disability was modified by physical activity and diet quality. The findings suggest that assessing comorbidity and promoting healthy lifestyles in clinical practice are important to improve stroke rehabilitation outcomes.

Supplementary Materials: The following are available online at https://www.mdpi.com/article/10.3390/nu13051641/s1, Table S1: Construct and convergent validity, internal consistency, floor and ceiling effects the healthy eating score ($n = 951$); Table S2: Associated factors of disability via bivariate linear regression analysis ($n = 951$).

Author Contributions: Conceptualization, L.T.K.N., B.N.D., D.N.V., K.M.P., M.-T.V., H.C.N., T.V.T., H.P.L., T.T.P.N., Q.M.N., C.Q.T., K.T.N., S.-H.Y., J.C.-J.C., and T.V.D.; data curation, L.T.K.N., B.N.D., D.N.V., M.-T.V., T.V.T., H.P.L., Q.M.N., and C.Q.T.; formal analysis, T.V.D.; funding acquisition, T.V.D.; investigation, L.T.K.N., B.N.D., D.N.V., K.M.P., M.-T.V., H.C.N., T.V.T., H.P.L., T.T.P.N., Q.M.N., C.Q.T., K.T.N., S.-H.Y., J.C.-J.C., and T.V.D.; methodology, L.T.K.N., B.N.D., D.N.V., K.M.P., M.-T.V., H.C.N., T.V.T., H.P.L., T.T.P.N., Q.M.N., C.Q.T., K.T.N., S.-H.Y., J.C.-J.C., and T.V.D.; project administration, T.T.P.N. and K.T.N.; resources, L.T.K.N., K.M.P., H.C.N., Q.M.N., C.Q.T., and T.V.D.; software, T.V.D.; supervision, L.T.K.N., K.M.P., S.-H.Y., J.C.-J.C., and T.V.D.; validation, L.T.K.N., B.N.D., D.N.V., K.M.P., M.-T.V., H.C.N., T.V.T., H.P.L., T.T.P.N., Q.M.N., C.Q.T., K.T.N., S.-H.Y., J.C.-J.C., and T.V.D.; writing—original draft, L.T.K.N. and T.V.D.; writing—review and editing, L.T.K.N., B.N.D., D.N.V., K.M.P., M.-T.V., H.C.N., T.V.T., H.P.L., T.T.P.N., Q.M.N., C.Q.T., K.T.N., S.-H.Y., J.C.-J.C., and T.V.D. All authors have read and agreed to the published version of the manuscript.

Funding: This research was supported by Taipei Medical University (108-6202-008-112).

Institutional Review Board Statement: The study was conducted according to the guidelines of the Declaration of Helsinki, and approved by the Ethics Committee of Hanoi University of Public Health (IRB No. 498/2019/YTCC-HD3 and No. 312/2020/YTCC-HD3).

Informed Consent Statement: Informed consent was obtained from all subjects involved in the study.

Data Availability Statement: Data will be available on the reasonable request from the corresponding author.

Acknowledgments: The authors would like to thank the doctors, nurses, and medical students who helped with data collection.

Conflicts of Interest: The authors declare no conflict of interest.

References

1. Sacco, R.L.; Kasner, S.E.; Broderick, J.P.; Caplan, L.R.; Connors, J.J.; Culebras, A.; Elkind, M.S.; George, M.G.; Hamdan, A.D.; Higashida, R.T.; et al. An updated definition of stroke for the 21st century: A statement for healthcare professionals from the American Heart Association/American Stroke Association. *Stroke* **2013**, *44*, 2064–2089. [CrossRef]
2. Feigin, V.L.; Krishnamurthi, R.V.; Parmar, P.; Norrving, B.; Mensah, G.A.; Bennett, D.A.; Barker-Collo, S.; Moran, A.E.; Sacco, R.L.; Truelsen, T.; et al. Update on the Global Burden of Ischemic and Hemorrhagic Stroke in 1990–2013: The GBD 2013 Study. *Neuroepidemiology* **2015**, *45*, 161–176. [CrossRef]

3. Yamanashi, H.; Ngoc, M.Q.; Van Huy, T.; Suzuki, M.; Tsujino, A.; Toizumi, M.; Takahashi, K.; Thiem, V.D.; Anh, D.D.; Anh, N.T.H.; et al. Population-Based Incidence Rates of First-Ever Stroke in Central Vietnam. *PLoS ONE* **2016**, *11*, e0160665. [CrossRef] [PubMed]

4. Pham, T.L.; Blizzard, L.; Srikanth, V.; Thrift, A.G.; Lien, N.T.; Thang, N.H.; Gall, S.L. Case-fatality and functional status three months after first-ever stroke in Vietnam. *J. Neurol. Sci.* **2016**, *365*, 65–71. [CrossRef]

5. Krishnamurthi, R.V.; Feigin, V.L.; Forouzanfar, M.H.; Mensah, G.A.; Connor, M.; Bennett, D.A.; Moran, A.E.; Sacco, R.L.; Anderson, L.M.; Truelsen, T.; et al. Global and regional burden of first-ever ischaemic and haemorrhagic stroke during 1990–2010: Findings from the Global Burden of Disease Study 2010. *Lancet Glob. Health* **2013**, *1*, e259–e281. [CrossRef]

6. Feigin, V.L.; Roth, G.A.; Naghavi, M.; Parmar, P.; Krishnamurthi, R.; Chugh, S.; Mensah, G.A.; Norrving, B.; Shiue, I.; Ng, M.; et al. Global burden of stroke and risk factors in 188 countries, during 1990–2013: A systematic analysis for the Global Burden of Disease Study 2013. *Lancet Neurol.* **2016**, *15*, 913–924. [CrossRef]

7. Benjamin, E.J.; Muntner, P.; Alonso, A.; Bittencourt, M.S.; Callaway, C.W.; Carson, A.P.; Chamberlain, A.M.; Chang, A.R.; Cheng, S.; Das, S.R.; et al. Heart disease and stroke statistics—2019 update: A report from the American heart association. *Circulation* **2019**, *139*, e56–e528. [CrossRef] [PubMed]

8. Gallacher, K.I.; Batty, G.D.; McLean, G.; Mercer, S.W.; Guthrie, B.; May, C.R.; Langhorne, P.; Mair, F.S. Stroke, multimorbidity and polypharmacy in a nationally representative sample of 1,424,378 patients in Scotland: Implications for treatment burden. *BMC Med.* **2014**, *12*, 151. [CrossRef]

9. Barnett, K.; Mercer, S.W.; Norbury, M.; Watt, G.; Wyke, S.; Guthrie, B. Epidemiology of multimorbidity and implications for health care, research, and medical education: A cross-sectional study. *Lancet* **2012**, *380*, 37–43. [CrossRef]

10. Ofori-Asenso, R.; Zomer, E.; Chin, K.L.; Si, S.; Markey, P.; Tacey, M.; Curtis, A.J.; Zoungas, S.; Liew, D. Effect of Comorbidity Assessed by the Charlson Comorbidity Index on the Length of Stay, Costs and Mortality among Older Adults Hospitalised for Acute Stroke. *Int. J. Environ. Res. Public Health* **2018**, *15*, 2532. [CrossRef]

11. Mohamed, W.; Bhattacharya, P.; Shankar, L.; Chaturvedi, S.; Madhavan, R. Which Comorbidities and Complications Predict Ischemic Stroke Recovery and Length of Stay? *Neurologist* **2015**, *20*, 27–32. [CrossRef]

12. Simić-Panić, D.; Bošković, K.; Milićević, M.; Žikić, T.R.; Bošnjak, M.C.; Tomašević-Todorović, S.; Jovićević, M. The Impact of Comorbidity on Rehabilitation Outcome after Ischemic Stroke. *Acta Clin. Croat.* **2018**, *57*, 5–15. [CrossRef] [PubMed]

13. Pandian, J.D.; Gall, S.L.; Kate, M.P.; Silva, G.S.; Akinyemi, R.O.; Ovbiagele, B.I.; Lavados, P.M.; Gandhi, D.B.C.; Thrift, A.G. Prevention of stroke: A global perspective. *Lancet* **2018**, *392*, 1269–1278. [CrossRef]

14. Salehi-Abargouei, A.; Maghsoudi, Z.; Shirani, F.; Azadbakht, L. Effects of Dietary Approaches to Stop Hypertension (DASH)-style diet on fatal or nonfatal cardiovascular diseases—Incidence: A systematic review and meta-analysis on observational prospective studies. *Nutrients* **2013**, *29*, 611–618. [CrossRef]

15. Struijk, E.A.; May, A.M.; Wezenbeek, N.L.; Fransen, H.P.; Soedamah-Muthu, S.S.; Geelen, A.; Boer, J.M.; van der Schouw, Y.T.; Bueno-De-Mesquita, H.B.; Beulens, J.W. Adherence to dietary guidelines and cardiovascular disease risk in the EPIC-NL cohort. *Int. J. Cardiol.* **2014**, *176*, 354–359. [CrossRef] [PubMed]

16. Paterson, K.E.; Myint, P.K.; Jennings, A.; Bain, L.K.; Lentjes, M.A.; Khaw, K.T.; Welch, A.A. Mediterranean Diet Reduces Risk of Incident Stroke in a Population With Varying Cardiovascular Disease Risk Profiles. *Stroke* **2018**, *49*, 2415–2420.

17. Lin, P.-H.; Yeh, W.-T.; Svetkey, L.P.; Chuang, S.-Y.; Chang, Y.-C.; Wang, C.; Pan, W.-H. Dietary intakes consistent with the DASH dietary pattern reduce blood pressure increase with age and risk for stroke in a Chinese population. *Asia Pac. J. Clin. Nutr.* **2013**, *22*, 482–491.

18. Chan, R.; Chan, D.; Woo, J. The association of a priori and a posterior dietary patterns with the risk of incident stroke in Chinese older people in Hong Kong. *J. Nutr. Health Aging* **2013**, *17*, 866–874. [CrossRef]

19. Virani, S.S.; Alonso, A.; Aparicio, H.J.; Benjamin, E.J.; Bittencourt, M.S.; Callaway, C.W.; Carson, A.P.; Chamberlain, A.M.; Cheng, S.; Delling, F.N.; et al. Heart Disease and Stroke Statistics-2021 Update: A Report From the American Heart Association. *Circulation* **2021**, *143*, e254–e743. [CrossRef]

20. Craig, C.L.; Marshall, A.L.; Sjöström, M.; Bauman, A.E.; Booth, M.L.; Ainsworth, B.E.; Pratt, M.; Ekelund, U.; Yngve, A.; Sallis, J.F.; et al. International physical activity questionnaire: 12-country reliability and validity. *Med. Sci. Sports Exerc.* **2003**, *35*, 1381–1395. [CrossRef] [PubMed]

21. Pham, T.; Bui, L.; Nguyen, A.; Nguyen, B.; Tran, P.; Vu, P.; Dang, L. The prevalence of depression and associated risk factors among medical students: An untold story in Vietnam. *PLoS ONE* **2019**, *14*, e0221432. [CrossRef]

22. Lee, P.H.; Macfarlane, D.J.; Lam, T.H.; Stewart, S.M. Validity of the international physical activity questionnaire short form (IPAQ-SF): A systematic review. *Int. J. Behav. Nutr. Phys. Act.* **2011**, *8*, 115. [CrossRef] [PubMed]

23. Quan, H.; Li, B.; Couris, C.M.; Fushimi, K.; Graham, P.; Hider, P.; Januel, J.-M.; Sundararajan, V. Updating and Validating the Charlson Comorbidity Index and Score for Risk Adjustment in Hospital Discharge Abstracts Using Data From 6 Countries. *Am. J. Epidemiol.* **2011**, *173*, 676–682. [CrossRef] [PubMed]

24. Charlson, M.E.; Pompei, P.; Ales, K.L.; MacKenzie, C. A new method of classifying prognostic comorbidity in longitudinal studies: Development and validation. *J. Chronic Dis.* **1987**, *40*, 373–383. [CrossRef]

25. Duong, T.V.; Aringazina, A.; Baisunova, G.; Nurjanah, N.; Pham, T.V.; Pham, K.M.; Truong, T.Q.; Nguyen, K.T.; Oo, W.M.; Su, T.T.; et al. Development and validation of a new short-form health literacy instrument (HLS-SF12) for the general public in six Asian countries. *Health Lit. Res. Pract.* **2019**, *3*, e91–e102. [CrossRef]

26. Van Duong, T.; Chiu, C.-H.; Lin, C.-Y.; Chen, Y.-C.; Wong, T.-C.; Chang, P.W.S.; Yang, S.-H. E-healthy diet literacy scale and its relationship with behaviors and health outcomes in Taiwan. *Health Promot. Int.* **2021**, *36*, 20–33. [CrossRef] [PubMed]
27. Van Duong, T.; Nguyen, T.T.P.; Pham, K.M.; Nguyen, K.T.; Giap, M.H.; Tran, T.D.X.; Nguyen, C.X.; Yang, S.-H.; Su, C.-T. Validation of the Short-Form Health Literacy Questionnaire (HLS-SF12) and Its Determinants among People Living in Rural Areas in Vietnam. *Int. J. Environ. Res. Public Health* **2019**, *16*, 3346. [CrossRef] [PubMed]
28. Ho, H.V.; Hoang, G.T.; Pham, V.T.; Duong, T.V.; Pham, K.M. Factors associated with health literacy among the elderly people in Vietnam. *BioMed Res. Int.* **2020**, *2020*, 1–7.
29. Nguyen, H.C.; Nguyen, M.H.; Do, B.N.; Tran, C.Q.; Nguyen, T.T.P.; Pham, K.M.; Pham, L.V.; Tran, K.V.; Duong, T.T.; Tran, T.V.; et al. People with Suspected COVID-19 Symptoms Were More Likely Depressed and Had Lower Health-Related Quality of Life: The Potential Benefit of Health Literacy. *J. Clin. Med.* **2020**, *9*, 965. [CrossRef] [PubMed]
30. Nguyen, H.; Do, B.; Pham, K.; Kim, G.; Dam, H.; Nguyen, T.; Nguyen, T.; Nguyen, Y.; Sørensen, K.; Pleasant, A.; et al. Fear of COVID-19 Scale—Associations of Its Scores with Health Literacy and Health-Related Behaviors among Medical Students. *Int. J. Environ. Res. Public Health* **2020**, *17*, 4164. [CrossRef] [PubMed]
31. HLS-EU Consortium. Comparative Report of Health Literacy in Eight EU Member States. The European Health Literacy Project 2009–2012. Available online: https://www.healthliteracyeurope.net/hls-eu (accessed on 22 October 2012).
32. Warren-Findlow, J.; Reeve, C.L.; Racine, E.F. Psychometric Validation of a Brief Self-report Measure of Diet Quality: The DASH-Q. *J. Nutr. Educ. Behav.* **2017**, *49*, 92–99.e1. [CrossRef]
33. World Health Organization. WHO Disability Assessment Schedule 2.0 (WHODAS 2.0). Available online: https://www.who.int/classifications/icf/whodasii/en/ (accessed on 15 October 2019).
34. Üstün, T.B.; Kostanjesek, N.; Chatterji, S.; Rehm, J. *Measuring Health and Disability: Manual for WHO Disability Assessment Schedule (WHODAS 2.0)*; World Health Organization: Geneva, Switzerland, 2010; 88p.
35. Üstün, T.B.; Chatterji, S.; Kostanjsek, N.; Rehm, J.; Kennedy, C.; Epping-Jordan, J.; Saxena, S.; Von Korff, M.; Pull, C. Developing the World Health Organization Disability Assessment Schedule 2.0. *Bull. World Health Organ.* **2010**, *88*, 815–823. [CrossRef]
36. Prime Minister of Vietnam. Gov't Extends Social Distancing for at Least One Week in 28 Localities. Available online: http://news.chinhphu.vn/Home/Govt-extends-social-distancing-for-at-least-one-week-in-28-localities/20204/39735.vgp (accessed on 15 April 2020).
37. Prime Minister of Vietnam. Gov't to Gradually Ease COVID-19 Control Measure in Cautious Manner. Available online: http://primeminister.chinhphu.vn/Home/Govt-to-gradually-ease-COVID19-control-measure-in-cautious-manner/20204/3759.vgp (accessed on 20 April 2020).
38. National Center for Immunization and Respiratory Diseases (NCIRD) Division of Viral Diseases. What Healthcare Personnel Should Know about Caring for Patients with Confirmed or Possible 2019-nCoV Infection. Available online: https://www.cdc.gov/coronavirus/2019-ncov/hcp/caring-for-patients.html (accessed on 7 February 2020).
39. World Health Organization (WHO). Country & Technical Guidance—Coronavirus Disease (COVID-19). Available online: https://www.who.int/emergencies/diseases/novel-coronavirus-2019/technical-guidance (accessed on 10 February 2020).
40. Ministry of Health. Coronavirus Disease (COVID-19) Outbreak in Vietnam. Available online: https://ncov.moh.gov.vn/ (accessed on 4 May 2020).
41. Hayes, A.F. *Introduction to Mediation, Moderation, and Conditional Process Analysis: A Regression-Based Approach*; Guilford Publications: New York, NY, USA, 2017; pp. 120–141.
42. IBM SPSS. *IBM SPSS Statistics for Windows*; Version 20.0; IBM Corp: New York, NY, USA, 2011.
43. Denti, L.; Artoni, A.; Casella, M.; Giambanco, F.; Scoditti, U.; Ceda, G.P. Validity of the Modified Charlson Comorbidity Index as Predictor of Short-term Outcome in Older Stroke Patients. *J. Stroke Cerebrovasc. Dis.* **2015**, *24*, 330–336. [CrossRef] [PubMed]
44. Caballero, P.E.J.; Espuela, F.L.; Cuenca, J.C.P.; Moreno, J.M.R.; Zamorano, J.D.P.; Naranjo, I.C. Charlson Comorbidity Index in Ischemic Stroke and Intracerebral Hemorrhage as Predictor of Mortality and Functional Outcome after 6 Months. *J. Stroke Cerebrovasc. Dis.* **2013**, *22*, e214–e218. [CrossRef] [PubMed]
45. Corraini, P.; Szépligeti, S.K.; Henderson, V.W.; Ording, A.G.; Horváth-Puhó, E.; Sørensen, H.T. Comorbidity and the increased mortality after hospitalization for stroke: A population-based cohort study. *J. Thromb. Haemost.* **2017**, *16*, 242–252. [CrossRef] [PubMed]
46. Rand, D.; Eng, J.J.; Tang, P.-F.; Hung, C.; Jeng, J.-S. Daily physical activity and its contribution to the health-related quality of life of ambulatory individuals with chronic stroke. *Health Qual. Life Outcomes* **2010**, *8*, 80. [CrossRef] [PubMed]
47. Ezeugwu, V.E.; Manns, P.J. Sleep Duration, Sedentary Behavior, Physical Activity, and Quality of Life after Inpatient Stroke Rehabilitation. *J. Stroke Cerebrovasc. Dis.* **2017**, *26*, 2004–2012. [CrossRef] [PubMed]
48. Paul, L.; Brewster, S.; Wyke, S.; Gill, J.M.R.; Alexander, G.; Dybus, A.; Rafferty, D. Physical activity profiles and sedentary behaviour in people following stroke: A cross-sectional study. *Disabil. Rehabil.* **2015**, *38*, 362–367. [CrossRef] [PubMed]
49. Wondergem, R.; Veenhof, C.; Wouters, E.M.; De Bie, R.A.; Visser-Meily, J.M.; Pisters, M.F. Movement Behavior Patterns in People With First-Ever Stroke. *Stroke* **2019**, *50*, 3553–3560. [CrossRef] [PubMed]
50. Kramer, S.F.; Hung, S.H.; Brodtmann, A. The Impact of Physical Activity Before and After Stroke on Stroke Risk and Recovery: A Narrative Review. *Curr. Neurol. Neurosci. Rep.* **2019**, *19*, 28. [CrossRef]

51. Peçanha, T.; Goessler, K.F.; Roschel, H.; Gualano, B. Social isolation during the COVID-19 pandemic can increase physical inactivity and the global burden of cardiovascular disease. *Am. J. Physiol. Circ. Physiol.* **2020**, *318*, H1441–H1446. [CrossRef] [PubMed]
52. Lippi, G.; Henry, B.M.; Sanchis-Gomar, F. Physical inactivity and cardiovascular disease at the time of coronavirus disease 2019 (COVID-19). *Eur. J. Prev. Cardiol.* **2020**, *27*, 906–908. [CrossRef] [PubMed]
53. Rees, K.; Takeda, A.; Martin, N.; Ellis, L.; Wijesekara, D.; Vepa, A.; Das, A.; Hartley, L.; Stranges, S. Mediterranean-style diet for the primary and secondary prevention of cardiovascular disease. *Cochrane Database Syst. Rev.* **2019**, *3*, CD009825. [CrossRef] [PubMed]
54. Chareonrungrueangchai, K.; Wongkawinwoot, K.; Anothaisintawee, T.; Reutrakul, S. Dietary Factors and Risks of Cardiovascular Diseases: An Umbrella Review. *Nutrients* **2020**, *12*, 1088. [CrossRef] [PubMed]
55. Chiavaroli, L.; Viguiliouk, E.; Nishi, S.K.; Blanco Mejia, S.; Rahelić, D.; Kahleová, H.; Salas-Salvadó, J.; Kendall, C.W.; Sievenpiper, J.L. DASH Dietary Pattern and Cardiometabolic Outcomes: An Umbrella Review of Systematic Reviews and Meta-Analyses. *Nutrients* **2019**, *11*, 338. [CrossRef]
56. Lagström, H.; Stenholm, S.; Akbaraly, T.; Pentti, J.; Vahtera, J.; Kivimäki, M.; Head, J. Diet quality as a predictor of cardiometabolic disease–free life expectancy: The Whitehall II cohort study. *Am. J. Clin. Nutr.* **2020**, *111*, 787–794. [CrossRef]
57. Glenn, A.J.; Viguiliouk, E.; Seider, M.; Boucher, B.A.; Khan, T.A.; Blanco Mejia, S.; Jenkins, D.J.A.; Kahleová, H.; Rahelić, D.; Salas-Salvadó, J.; et al. Relation of Vegetarian Dietary Patterns With Major Cardiovascular Outcomes: A Systematic Review and Meta-Analysis of Prospective Cohort Studies. *Front. Nutr.* **2019**, *6*, 80. [CrossRef]
58. Sakai, K.; Kinoshita, S.; Tsuboi, M.; Fukui, R.; Momosaki, R.; Wakabayashi, H. Effects of Nutrition Therapy in Older Stroke Patients Undergoing Rehabilitation: A Systematic Review and Meta-Analysis. *J. Nutr. Health Aging* **2019**, *23*, 21–26. [CrossRef]
59. Agarwal, P.; Wang, Y.; Buchman, A.S.; Bennett, D.A.; Morris, M.C. Dietary Patterns and Self-reported Incident Disability in Older Adults. *J. Gerontol. Ser. A Boil. Sci. Med. Sci.* **2018**, *74*, 1331–1337. [CrossRef]
60. Tanaka, T.; Talegawkar, S.; Jin, Y.; Bandinelli, S.; Ferrucci, L. Association of Adherence to the Mediterranean-Style Diet with Lower Frailty Index in Older Adults. *Nutrients* **2021**, *13*, 1129. [CrossRef]
61. Huang, C.H.; Martins, B.A.; Okada, K.; Matsushita, E.; Uno, C.; Satake, S.; Kuzuya, M. A 3-year prospective cohort study of dietary patterns and frailty risk among community-dwelling older adults. *Clin. Nutr.* **2021**, *40*, 229–236. [CrossRef]
62. Huang, C.H.; Okada, K.; Matsushita, E.; Uno, C.; Satake, S.; Martins, B.A.; Kuzuya, M. Dietary Patterns and Muscle Mass, Muscle Strength, and Physical Performance in the Elderly: A 3-Year Cohort Study. *J. Nutr. Health Aging* **2021**, *25*, 108–115. [CrossRef]
63. Sato, Y.; Yoshimura, Y.; Abe, T. Nutrition in the First Week after Stroke Is Associated with Discharge to Home. *Nutrients* **2021**, *13*, 943. [CrossRef]
64. Bonaccio, M.; Pounis, G.; Cerletti, C.; Donati, M.B.; Iacoviello, L.; De Gaetano, G.; MOLI-SANI Study Investigators. Mediterranean diet, dietary polyphenols and low grade inflammation: Results from the MOLI-SANI study. *Br. J. Clin. Pharmacol.* **2016**, *83*, 107–113. [CrossRef] [PubMed]
65. Zabetakis, I.; Lordan, R.; Norton, C.; Tsoupras, A. COVID-19: The Inflammation Link and the Role of Nutrition in Potential Mitigation. *Nutrients* **2020**, *12*, 1466. [CrossRef]
66. Detopoulou, P.; Demopoulos, C.; Antonopoulou, S. Micronutrients, Phytochemicals and Mediterranean Diet: A Potential Protective Role against COVID-19 through Modulation of PAF Actions and Metabolism. *Nutrients* **2021**, *13*, 462. [CrossRef] [PubMed]
67. Tsoupras, A.; Lordan, R.; Zabetakis, I. Thrombosis and COVID-19: The Potential Role of Nutrition. *Front. Nutr.* **2020**, *7*, 583080. [CrossRef]
68. Fraiman, P.; Junior, C.G.; Moro, E.; Cavallieri, F.; Zedde, M. COVID-19 and Cerebrovascular Diseases: A Systematic Review and Perspectives for Stroke Management. *Front. Neurol.* **2020**, *11*, 574694. [CrossRef]
69. Oliveira, D.M.G.; Aguiar, L.T.; Limones, M.V.D.O.; Gomes, A.G.; Da Silva, L.C.; Faria, C.D.C.D.M.; Scalzo, P.L. Aerobic Training Efficacy in Inflammation, Neurotrophins, and Function in Chronic Stroke Persons: A Randomized Controlled Trial Protocol. *J. Stroke Cerebrovasc. Dis.* **2019**, *28*, 418–424. [CrossRef] [PubMed]
70. Friedman, E.; Shorey, C. Inflammation in multimorbidity and disability: An integrative review. *Health Psychol.* **2019**, *38*, 791–801. [CrossRef]
71. Yoshimura, Y.; Bise, T.; Nagano, F.; Shimazu, S.; Shiraishi, A.; Yamaga, M.; Koga, H. Systemic Inflammation in the Recovery Stage of Stroke: Its Association with Sarcopenia and Poor Functional Rehabilitation Outcomes. *Prog. Rehabil. Med.* **2018**, *3*, 20180011. [CrossRef] [PubMed]
72. Brummel, N.E.; Hughes, C.G.; Thompson, J.L.; Jackson, J.C.; Pandharipande, P.; McNeil, J.B.; Raman, R.; Orun, O.M.; Ware, L.B.; Bernard, G.R.; et al. Inflammation and Coagulation during Critical Illness and Long-Term Cognitive Impairment and Disability. *Am. J. Respir. Crit. Care Med.* **2021**, *203*, 699–706. [CrossRef] [PubMed]
73. May, A.M.; Struijk, E.A.; Fransen, H.P.; Onland-Moret, N.C.; De Wit, G.A.; Boer, J.M.A.; Van Der Schouw, Y.T.; Hoekstra, J.; Bueno-De-Mesquita, H.B.; Peeters, P.H.M.; et al. The impact of a healthy lifestyle on Disability-Adjusted Life Years: A prospective cohort study. *BMC Med.* **2015**, *13*, 1–9. [CrossRef]
74. Sánchez-Sánchez, E.; Ramírez-Vargas, G.; Avellaneda-López, Y.; Orellana-Pecino, J.I.; García-Marín, E.; Díaz-Jimenez, J. Eating Habits and Physical Activity of the Spanish Population during the COVID-19 Pandemic Period. *Nutrients* **2020**, *12*, 2826. [CrossRef] [PubMed]

75. Rodríguez-Pérez, C.; Molina-Montes, E.; Verardo, V.; Artacho, R.; García-Villanova, B.; Guerra-Hernández, E.J.; Ruíz-López, M.D. Changes in Dietary Behaviours during the COVID-19 Outbreak Confinement in the Spanish COVIDiet Study. *Nutrients* **2020**, *12*, 1730. [CrossRef]
76. Ismail, L.C.; Osaili, T.M.; Mohamad, M.N.; Al Marzouqi, A.; Jarrar, A.H.; Abu Jamous, D.O.; Magriplis, E.; Ali, H.I.; Al Sabbah, H.; Hasan, H.; et al. Eating Habits and Lifestyle during COVID-19 Lockdown in the United Arab Emirates: A Cross-Sectional Study. *Nutrients* **2020**, *12*, 3314. [CrossRef]
77. Niewada, M.; Michel, P. Lifestyle modification for stroke prevention: Facts and fiction. *Curr. Opin. Neurol.* **2016**, *29*, 9–13. [CrossRef]
78. Arnett, D.K.; Blumenthal, R.S.; Albert, M.A.; Buroker, A.B.; Goldberger, Z.D.; Hahn, E.J.; Himmelfarb, C.D.; Khera, A.; Lloyd-Jones, D.; McEvoy, J.W. 2019 ACC/AHA Guideline on the Primary Prevention of Cardiovascular Disease: A Report of the American College of Cardiology/American Heart Association Task Force on Clinical Practice Guidelines. *Circulation* **2019**, *140*, e596–e646. [CrossRef]

nutrients

Article

Possible Neuroprotective Effects of L-Carnitine on White-Matter Microstructural Damage and Cognitive Decline in Hemodialysis Patients

Yuji Ueno [1,*], Asami Saito [2,3], Junichiro Nakata [4], Koji Kamagata [2], Daisuke Taniguchi [1], Yumiko Motoi [1], Hiroaki Io [5], Christina Andica [2], Atsuhiko Shindo [1], Kenta Shiina [1], Nobukazu Miyamoto [1], Kazuo Yamashiro [6], Takao Urabe [6], Yusuke Suzuki [4], Shigeki Aoki [2] and Nobutaka Hattori [1]

[1] Department of Neurology, Juntendo University Faculty of Medicine, Tokyo 113-8421, Japan; dtanigu@juntendo.ac.jp (D.T.); motoi@juntendo.ac.jp (Y.M.); at-shindo@juntendo.ac.jp (A.S.); k-shiina@juntendo.ac.jp (K.S.); nobu-m@juntendo.ac.jp (N.M.); nhattori@juntendo.ac.jp (N.H.)

[2] Department of Radiology, Juntendo University Faculty of Medicine, Tokyo 113-8421, Japan; nqg16274@nifty.com (A.S.); kkamagat@juntendo.ac.jp (K.K.); andicach@gmail.com (C.A.); saoki@juntendo.ac.jp (S.A.)

[3] Department of Neurology and Stroke Medicine, Graduate School of Medicine, Yokohama City University, Yokohama 236-0004, Japan

[4] Department of Nephrology, Juntendo University Faculty of Medicine, Tokyo 113-8421, Japan; jnakata@juntendo.ac.jp (J.N.); yusuke@juntendo.ac.jp (Y.S.)

[5] Department of Nephrology, Juntendo University Nerima Hospital, Tokyo 177-8521, Japan; hiroaki@juntendo.ac.jp

[6] Department of Neurology, Juntendo University Urayasu Hospital, Urayasu 279-0021, Japan; kazuo-y@juntendo.ac.jp (K.Y.); t_urabe@juntendo.ac.jp (T.U.)

* Correspondence: yuji-u@juntendo.ac.jp; Tel.: +81-3-3813-3111; Fax: +81-3-5800-0547

Citation: Ueno, Y.; Saito, A.; Nakata, J.; Kamagata, K.; Taniguchi, D.; Motoi, Y.; Io, H.; Andica, C.; Shindo, A.; Shiina, K.; et al. Possible Neuroprotective Effects of L-Carnitine on White-Matter Microstructural Damage and Cognitive Decline in Hemodialysis Patients. *Nutrients* **2021**, *13*, 1292. https://doi.org/10.3390/nu13041292

Academic Editor: Yoshihiro Yoshimura

Received: 9 March 2021
Accepted: 12 April 2021
Published: 14 April 2021

Publisher's Note: MDPI stays neutral with regard to jurisdictional claims in published maps and institutional affiliations.

Abstract: Although L-carnitine alleviated white-matter lesions in an experimental study, the treatment effects of L-carnitine on white-matter microstructural damage and cognitive decline in hemodialysis patients are unknown. Using novel diffusion magnetic resonance imaging (dMRI) techniques, white-matter microstructural changes together with cognitive decline in hemodialysis patients and the effects of L-carnitine on such disorders were investigated. Fourteen hemodialysis patients underwent dMRI and laboratory and neuropsychological tests, which were compared across seven patients each in two groups according to duration of L-carnitine treatment: (1) no or short-term L-carnitine treatment (NSTLC), and (2) long-term L-carnitine treatment (LTLC). Ten age- and sex-matched controls were enrolled. Compared to controls, microstructural disorders of white matter were widely detected on dMRI of patients. An autopsy study of one patient in the NSTLC group showed rarefaction of myelinated fibers in white matter. With LTLC, microstructural damage on dMRI was alleviated along with lower levels of high-sensitivity C-reactive protein and substantial increases in carnitine levels. The LTLC group showed better achievement on trail making test A, which was correlated with amelioration of disorders in some white-matter tracts. Novel dMRI tractography detected abnormalities of white-matter tracts after hemodialysis. Long-term treatment with L-carnitine might alleviate white-matter microstructural damage and cognitive impairment in hemodialysis patients.

Keywords: L-carnitine; hemodialysis; vascular dementia; diffusion tensor imaging; diffusion kurtosis imaging; neurite orientation dispersion and density imaging

1. Introduction

Correlated with the substantial increase in patients with end-stage renal disease due to diabetes mellitus, hypertension, and obesity, the number of patients on maintenance dialysis has risen greatly, to 284 individuals per million population worldwide [1]. Cognitive impairment is critical in hemodialysis patients and is associated with death and dialysis withdrawal [2]. A previous large-scale cohort study of hemodialysis patients documented the prevalence of diagnosed dementia as 4%, whereas several studies investigated cognitive function tests of hemodialysis patients, showing cognitive impairment in as many as 87% [2–4]. Thus, there may be more undiagnosed or covert vascular dementia in hemodialysis patients than expected.

Diffusion tensor imaging (DTI) and diffusion kurtosis imaging (DKI) allow noninvasive investigation of the neural architecture of the brain using a Gaussian and non-Gaussian model, respectively [5,6]. Furthermore, multi-shell diffusion MRI techniques (msdMRI), including the neurite orientation dispersion and density imaging (NODDI) model consisting of intracellular volume fraction (ICVF), orientation dispersion index (ODI), and isotropic volume fraction (ISO), have been shown to more sensitively evaluate neuritic microstructure alterations than DTI [7]. Using DTI, the association of white-matter fiber tract disorders with cognitive decline has been reported in vascular dementia, as well as small vessel diseases in hemodialysis patients [8–11], although no evidence is currently available for the analysis of white-matter fiber tracts using msdMRI techniques.

In experimental studies, we recently created rat models of vascular dementia by ligation of bilateral common carotid arteries, creating chronic cerebral hypoperfusion, which induced cerebral white-matter damage and vascular dementia [12,13]. L-Carnitine was shown to play a pivotal role in suppressing inflammation, oxidative stress, and apoptosis in chronic debilitating diseases [14]. L-Carnitine also improves cardiac dysfunction, as well as peripheral artery disease [15,16]. Importantly, continuous supplementation with L-carnitine for a longer period alleviated oxidative stress in oligodendrocytes and promoted axonal outgrowth and myelination in cerebral white matter, thereby improving cognitive impairment after chronic cerebral hypoperfusion in rats [13]. So far, no specific therapeutic agents for vascular cognitive dysfunction associated with cerebral white-matter lesions are available in clinical practice.

Cognitive dysfunction in hemodialysis patients can be induced by a disorder of cerebral circulation due to atherosclerosis in cerebral arteries, oxidative stress, inflammation, and possibly blood pressure fluctuation due to hemodialysis, which might share common pathological mechanisms with chronic cerebral hypoperfusion in our experimental studies [12,13]. Taking advantage of the utility of novel dMRI techniques, we sought to assess the alterations of the microstructure of white-matter tracts in hemodialysis patients. Importantly, a postmortem brain study of a hemodialysis patient enrolled in the present study was carried out. Meanwhile, several studies with small sample sizes of 10 to 17 participants were conducted to explore the pathophysiology of microstructural damage of white-matter tracts using NODDI [17,18]. L-Carnitine alleviated white-matter lesions and cognitive impairment in an experimental study, but its effects on cognitive decline in hemodialysis patients are essentially unknown. The aim of the current study was to elucidate the therapeutic efficacy of L-carnitine for disorders of white-matter tracts seen on dMRI and cognitive decline in hemodialysis patients.

2. Materials and Methods

2.1. Study Participants

Patients were included on the basis of the following inclusion criteria, (i) age between 55 and 90 years, (ii) undergoing hemodialysis in Juntendo University Hospital Dialysis Center; and (iii) having at least one atherosclerotic vascular risk factor. Exclusion criteria were as follows: (i) any contraindication to brain MRI, (ii) moderate to severe cognitive decline or aphasia (need some assistance to complete their daily living activities or communication), (iii) coexistence of neurological diseases, and (iv) presence of severe pneumonia, chronic heart failure, or advanced cancer. Previous studies with sample sizes of 10 to 17 participants were conducted to explore the pathophysiology of microstructural alterations of white-matter tracts using NODDI [17,18]. Thus, 14 maintenance hemodialysis patients (age 70.8 ± 7.2 years, six women and eight men), treated three times a week at outpatient hemodialysis units in Juntendo University Hospital, were included in the present study. In addition, 10 age- and sex-matched healthy controls (73.0 ± 4.4 years, five women and five men) who were literate or able to be interviewed and could communicate effectively, with no history of severe mental disorders or dementia, were also recruited. This study was conducted in accordance with the Declaration of Helsinki. The independent ethics committee of Juntendo University Hospital approved this study (16–170). All study subjects were given an explanation of the study, and written, informed consent was obtained for the study objective, MRI and cognitive function tests, enrolment in the study, and ensuring confidentiality of information.

2.2. Risk Factors

Atherosclerotic vascular risk factors were defined according to the previous literature [19]: (1) hypertension, history of using antihypertensive agents, systolic blood pressure >140 mmHg, or diastolic blood pressure >90 mmHg; (2) diabetes mellitus, use of oral hypoglycemic agents or insulin, or glycosylated hemoglobin >6.4%; (3) dyslipidemia, use of antihyperlipidemic agents, serum LDL-C ≥140 mg/dL, HDL-C <40 mg/dL, or triglycerides ≥150 mg/dL; (4) current smoking status; (5) coronary artery disease, previous history of angina pectoris or myocardial infarction; (6) stroke, previous history of ischemic stroke, hemorrhagic stroke, or undetermined stroke.

2.3. Drug Administration and Classification

The protocol for treatment with L-carnitine (Otsuka Pharmaceutical Co., Ltd., Tokushima, Japan) was 600 mg orally once a day starting in July 2012, followed by 1000 mg intravenously per hemodialysis day starting in January 2014. Termination of L-carnitine treatment was determined randomly, and patients without L-carnitine treatment were also included. On the basis of the duration of L-carnitine treatment, seven patients each were classified into the no or short-term L-carnitine treatment (NSTLC) and LTLC long-term L-carnitine treatment (LTLC) groups, with an allocation ratio of 1:1. In the NSTLC group, four patients' intravenous L-carnitine treatment was stopped from June 2014 to April 2015; thus, L-carnitine was discontinued on the day of dMRI and neuropsychological tests (January 2017 to September 2017), whereas the remaining three patients were not treated with L-carnitine. In the LTLC group, intravenous L-carnitine treatment was continued on the day of dMRI and neuropsychological tests (January 2017 to September 2017) in seven patients.

2.4. Profile of Carnitine Kinetics after L-Carnitine Treatment and Other Laboratory Blood Examinations

Serum total, free, and acylated carnitine concentrations were measured using the enzyme cycling method, and the ratio of acylated carnitine to free carnitine was calculated. According to the above treatment protocol of L-carnitine, total, free, and acylated carnitine concentrations were measured before treatment, at 6 months after 600 mg of L-carnitine treatment orally per day, and at 6 months after 1000 mg of L-carnitine treatment intravenously per hemodialysis day. Other laboratory data (calcium, phosphate, glucose, low-density lipoprotein, high-density lipoprotein, hemoglobin, triglycerides, creatinine, eGFR, leukocyte count, and high-sensitivity C-reactive protein) were examined by standard enzymatic methods. The averages of these data in three consecutive tests before MRI were calculated.

2.5. Neuropsychological Tests

The mini mental state examination (MMSE), Hasegawa dementia rating scale-revised (HDS-R) for standard cognitive function, the frontal assessment battery (FAB), and the Japanese version of the Montreal cognitive assessment (MoCA-J) were performed. The trail making test (TMT) was also conducted, consisting of two parts. TMT part A is a task of rapid visual scanning and processing speed, in which the respondent connects randomly arranged numbers in consecutive order. In part B, the respondent connects randomly arranged letters and numbers in consecutive order, alternating between numbers and letters. The four subtests of the Wechsler Memory Scale-Revised (WMS-R), namely, logical memory I/II, digit span, and visual span, were also carried out. All neuropsychological tests were performed by an experienced neuropsychologist (MS) within 60 min for each patient. The standard scores of TMT and four subsets of WMS-R were calculated by average score and standard deviation for each age group from previous data of Japanese subjects (Supplementary Table S1) [20], using the formula Z (standard score) = x (raw score) − μ (mean score)/σ (standard deviation).

2.6. MRI Acquisition

MRI data were acquired on a 3 T MRI scanner (MAGNETOM Prisma; Siemens Healthcare, Erlangen, Germany) with a 64-channel head coil. Sequences of conventional MRI included fluid-attenuated inversion recovery imaging (FLAIR) and the GRE T2 * sequence. FLAIR (TR/TE = 10,000/114 ms) was used to evaluate the degree of deep and subcortical white-matter hyperintensity (DSWMH) and periventricular hyperintensity (PVH) in accordance with Fazekas grades 0–3 [21], and GRE T2 * (TR/TE = 410–500/12 ms) was used to identify cerebral microbleeds (CMBs), defined as rounded areas of signal loss with diameter <10 mm. Diffusion-weighted imaging (DWI) was performed by spin-echo echo-planar imaging in the anterior-to-posterior phase-encoding direction. Multi-shell DWI was obtained with b-values of 1000 and 2000 s/mm^2 along 64 isotropic diffusion gradients for each shell. Each DWI acquisition was completed with a gradient-free image (b = 0). The acquisition parameters were as follows: repetition time/echo time = 3300/70 ms/ms, field of view = 229 × 229 mm, matrix = 130 × 130, resolution = 1.8 × 1.8 mm, slice thickness = 1.8 mm, and acquisition time = 7.29 min.

2.7. MRI Preprocessing

The diffusion datasets were corrected for eddy-current distortions and small head movements and denoised using EDDY and TOPUP toolboxes [22]. The resulting images were visually checked in axial, coronal, and sagittal views. All images were free of severe artefacts, such as gross geometric distortion, signal dropout, or bulk motion. The resulting images were used to acquire (1) fractional anisotropy (FA), MD (mean diffusivity), AD (axial diffusivity), and RD (radial diffusivity) maps using ordinary least squares applied to the diffusion-weighted images with b = 0 and 1000 s/mm^2, (2) mean kurtosis (MK), axial kurtosis (AK), and radial kurtosis (RK) using Diffusion Kurtosis Estimator (https://www.nitrc.org/projects/dke/, accessed on 1 June 2018), and (3) ICVF, ODI, and ISO maps using the NODDI model implemented in NODDI Matlab Toolbox5 (http://www.nitrc.org/projects/noddi_toolbox, accessed on 1 April 2019) and AMICO.

2.8. Tract-Based Spatial Statistics Analysis

Tract-based spatial statistics (TBSS), a voxel-wise statistical analysis of whole-brain white matter, was implemented in FMRIB Software Library version 5.0.9 (FSL; Oxford Centre for Functional MRI of the Brain, Oxford, UK; www.fmrib.ox.ac.uk/fsl, accessed on 1 September 2019) [21]. TBSS was performed to identify significant differences between groups of each of the DTI, DKI, and NODDI indices.

2.9. Tract-of-Interest (TOI) Analysis

Any maps showing significant clusters by TBSS analysis were localized using the Johns Hopkins University's ICBM-DTI-81 WM tractography atlas, which is composed of the left (right) anterior thalamic radiation (L[R]atr), corticospinal tract (L[R]cs), cingulum cingulate gyrus (L[R]cc), cingulum hippocampus (L[R]ch), forceps major (fm), forceps minor (fmi), inferior fronto-occipital fasciculus (L[R]ifof), inferior longitudinal fasciculus (L[R]ilf), superior longitudinal fasciculus (L[R]slf), uncinate fasciculus (L[R]uf), and slf temporal part (L[R]slftemp). The diffusion measures were averaged in each TOI delineated by the atlas. Correlation analysis was performed to explore the relationships between each index and neuropsychological scores [23].

2.10. Histopathological Analysis of an Autopsy Brain

An autopsy brain from a patient was fixed with 15% neutral buffered formalin, and the selected sections were embedded in paraffin. The paraffin-embedded blocks were sliced at a thickness of 6 μm. Brain sections were stained with hematoxylin and eosin (H&E), Klüber–Barrera (KB), and immunohistochemical stains using several antibodies for specific proteins. For immunohistochemistry, brain sections underwent antigen retrieval by heat activation in an autoclave with or without formic acid pretreatment, before being incubated overnight at 4 °C in primary antibody. The primary antibodies used were against phosphorylated tau, amyloid beta, and neurofilament heavy weight (NFH). Bound antibodies were visualized using the peroxidase-polymer-based method using a Histofine Simple Stain MAX-PO kit (Nichirei, Tokyo, Japan) with diaminobenzidine as the chromogen.

2.11. Statistical Analysis

Values presented in this study are expressed as means ± standard deviation. All statistical analyses were performed using IBM SPSS Statistics for Windows, version 22.0 (IBM Corporation, Armonk, NY, USA), except for general linear model (GLM) analysis, which was performed using the FSL. The Shapiro–Wilk test was used to assess data normality, whereas demographic data were analyzed using the unpaired Student's *t*-test for continuous variables, and the chi-squared test was used for categorical variables. The effect size was then calculated using Cohen's *d* to evaluate the statistical power of the relationship determined during group comparisons [24]. Clinical features, values on psychological tests, laboratory data, and degree of PVH and DSWMH were analyzed using the Mann–Whitney test for nonparametric variables and the chi-squared test for categorical variables between hemodialysis patients in the NSTLC and LTLC groups. For TBSS analysis, a GLM framework with one-way analysis of variance (ANOVA; healthy controls vs. hemodialysis patients with NSTLC vs. hemodialysis patients with LTLC) with age and sex as covariates were applied in the FSL randomize tool with 5000 permutations. The results were then corrected for multiple comparisons by controlling family-wise error and applying the threshold-free cluster enhancement option. For TOI analysis, one-way ANOVA with the Welch and Games–Howell post hoc tests was performed to compare among the three groups. Spearman's rank correlation coefficient was used to test for any linear associations between the diffusion metrics and psychological test scores that reached significant differences between the NSTLC and LTLC groups. The false discovery rate (FDR) was used to correct for multiple comparisons (20 TOIs). In all analyses, a probability value <0.05 was considered significant.

3. Results

3.1. Study Population

The baseline characteristics of the 14 enrolled patients are summarized in Table 1; 93%, 43%, and 57% of the patients had hypertension, diabetes mellitus, and dyslipidemia, respectively, and 14% and 21% of patients had coronary artery and peripheral artery diseases, respectively. Causes of hemodialysis were diabetic nephropathy in six patients, chronic glomerulonephritis in three patients, and nephrosclerosis, polycystic kidney disease, rapidly progressive glomerulonephritis, and renal cell carcinoma in one patient each.

Table 1. Baseline characteristics, laboratory data, and cognitive and radiological findings according to total duration of L-carnitine treatment in hemodialysis patients.

		Duration of LCAR Treatment		*p*
	Total	No or Short-Term	Long-Term	
Characteristic	*n* = 14	*n* = 7	*n* = 7	
Sociodemographic				
Age, years, mean ± SD	70.8 ± 7.2	72.6 ± 9.7	69.0 ± 3.3	0.259
Sex, male, no. (%)	8 (57)	4 (57)	4 (57)	0.589
Final education				0.431
High school	4 (29)	1 (14)	3 (43)	
Business school	2 (14)	1 (14)	1 (14)	
Junior college	3 (21)	1 (14)	2 (29)	
University	5 (36)	4 (57)	1 (14)	
Body mass index	20.6 ± 4.2	21.8 ± 5.3	19.3 ± 2.3	0.456
Risk factors, no. (%)				
Hypertension	13 (93)	7 (100)	6 (86)	1
Diabetes mellitus	6 (43)	3 (43)	3 (43)	0.589
Dyslipidemia	8 (57)	3 (43)	5 (71)	0.589
Current cigarette smoking	1 (7)	1 (14)	0 (0)	1
Coronary artery disease	2 (14)	1 (14)	1 (14)	0.445
Previous history of stroke	3 (21)	2 (29)	1 (14)	1
Cause of renal failure				0.718
Glomerulonephritis	3 (21)	1 (14)	2 (29)	
Diabetic nephropathy	6 (43)	3 (43)	3 (43)	
Nephrosclerosis	1 (7)	1 (14)	0 (0)	
Others	4 (29)	2 (29)	2 (29)	
Total time of hemodialysis	1217 ± 687	870 ± 750	1564 ± 405	0.097
Total amount of LCAR, g	496.4 ± 311.4	241.3 ± 233.1	751.4 ± 62.6	0.002
Laboratory findings, mean ± SD				
Leukocyte count, 102/μL	60.9 ± 16.6	64.1 ± 19.1	57.7 ± 13.4	0.413
LDL-C, mg/dL	84.5 ± 24.1	91.0 ± 28.7	78.0 ± 16.6	0.195
HDL-C, mg/dL	55.5 ± 15.1	51.0 ± 13.9	60.1 ± 15.1	0.082
Triglycerides, mg/dL	109.3 ± 43.7	112.2 ± 43.4	106.3 ± 43.4	0.772
Glucose, mg/dL	133.9 ± 45.5	135.4 ± 44.8	132.4 ± 47.2	0.782
eGFR, mL/min	4.7 ± 2.5	4.3 ± 1.8	5.2 ± 3.1	0.481
Creatinine, mg/dL	10.1 ± 2.6	10.2 ± 3.2	9.9 ± 2.0	0.93
Hs-CRP, mg/dL	0.41 ± 0.65	0.65 ± 0.68	0.17 ± 0.52	<0.001
Calcium, mg/dL	9.0 ± 0.6	8.8 ± 0.8	9.2 ± 0.2	0.173
Phosphate, mg/dL	5.2 ± 1.4	5.5 ± 1.5	5.0 ± 1.3	0.191
Conventional MRI				
PVH, grade 0–3	1.5 ± 0.7	1.4 ± 0.5	1.6 ± 0.8	0.902
DSWMH, grade 0–3	1.1 ± 0.7	1.1 ± 0.7	1.0 ± 0.8	0.805
Number of CMBs	2.9 ± 6.4	5.1 ± 8.7	0.7 ± 1.0	0.318
Cognitive function test				
MMSE	25.4 ± 3.5	26.0 ± 2.6	24.9 ± 4.3	0.71
HDS-R	26.3 ± 2.3	27.1 ± 1.5	25.4 ± 2.8	0.259
FAB	14.9 ± 3.2	15.0 ± 4.2	14.7 ± 2.0	0.318
MoCA-J	23.9 ± 3.2	23.9 ± 3.8	23.9 ± 2.8	0.71

The chi-squared test and the Mann–Whitney U test were used for comparisons. LCAR = L-carnitine; LDL-C = low-density lipoprotein cholesterol; HDL-C = high-density lipoprotein cholesterol; eGFR = estimated glomerular filtration rate; hs-CRP = high-sensitivity C-reactive protein. MRI = magnetic resonance imaging; PVH = periventricular hyperintensity; DSWMH = deep and subcortical white matter hyperintensity; MMSE = mini mental state examination; HDS-R = Hasegawa dementia rating scale-revised; FAB = frontal assessment battery; MoCA-J = Montreal cognitive assessment Japanese version.

3.2. Profile of Serum Carnitine Kinetics before and after L-Carnitine Treatment and Classification

In 11 patients, treatment with 600 mg of L-carnitine orally per day was started in July 2012, followed by 1000 mg of L-carnitine intravenously per hemodialysis day from January 2014, whereas the remaining three patients were not treated with L-carnitine. The total amount of L-carnitine was significantly higher in the LTLC group than in the NSTLC group (751.4 ± 62.6 g vs. 241.3 ± 233.1 g, p = 0.002, Table 1). In the 11 patients who were treated with L-carnitine, serum levels of total carnitine were 47.4 ± 12.1 μmol/L at baseline, increased to 195.0 ± 57.2 μmol/L at 6 months after 600 mg of oral L-carnitine treatment (p < 0.01), and further increased substantially to 548.9 ± 148.2 μmol/L at 6 months after 1000 mg of intravenous L-carnitine treatment (p < 0.001, vs. baseline and after oral L-carnitine treatment, respectively, Figure 1). Free and acylated carnitine levels were

29.7 ± 10.3 µmol/L and 17.7 ± 3.1 µmol/L at baseline, increased to 123.8 ± 40.6 µmol/L and 71.2 ± 23.1 µmol/L after 600 mg of oral L-carnitine treatment ($p < 0.01$ and $p < 0.05$, respectively), and further increased to 332.2 ± 84.3 µmol/L and 216.8 ± 74.5 µmol/L after 1000 mg of intravenous L-carnitine treatment, respectively ($p < 0.001$, vs. baseline and after oral L-carnitine treatment, respectively, Figure 1). The ratio of acylated carnitine to free carnitine did not differ after L-carnitine treatment.

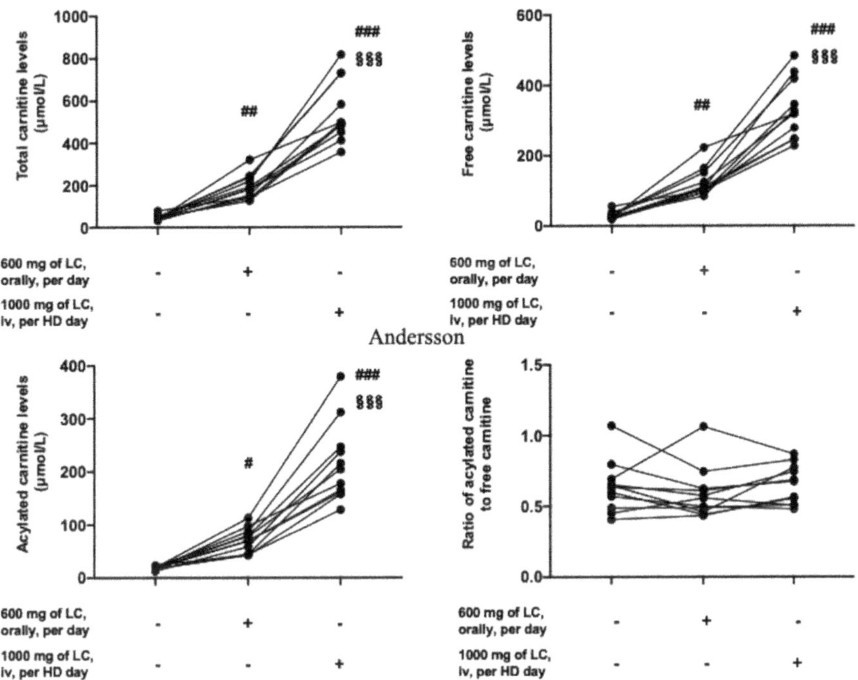

Figure 1. Temporal profile of carnitine kinetics after L-carnitine treatment. Serum total, free, and acylated carnitine concentrations were measured, and the ratios of acylated carnitine to free carnitine were calculated before L-carnitine treatment, at 6 months after 600 mg of L-carnitine treatment orally a day, and at 6 months after 1000 mg of L-carnitine treatment per hemodialysis day. # $p < 0.05$, ## $p < 0.01$, ### $p < 0.001$ vs. at baseline; §§§ $p < 0.001$, vs. at 6 months after 600 mg of L-carnitine treatment orally. LC = L-carnitine; HD = hemodialysis.

3.3. Baseline Patients' Profile, Degree of White-Matter Lesions, Psychological Tests, and Laboratory Data in the NSTLC and LTLC Groups

Between the NSTLC and LTLC groups, age, frequency of male sex, final education level, atherosclerotic risk factors, vascular diseases, and cause of renal failure were not significantly different (Table 1). On laboratory data, high-sensitivity C-reactive protein (hs-CRP) levels were significantly lower in hemodialysis patients with LTLC than in those with NSTLC (0.17 ± 0.52 vs. 0.65 ± 0.68 mg/dL, $p < 0.001$), whereas other laboratory markers did not show any significant differences. Degrees of PVH and DSWMH and number of CMBs were not different between groups.

3.4. Cognitive Functions in Hemodialysis Patients with NSTLC and LTLC

Overall, MMSE, HDS-R, FAB, and MoCA-J scores were 25.4 ± 3.5, 26.3 ± 2.3, 14.9 ± 3.2, and 23.9 ± 3.2 points, respectively, which were not different between the NSTLC and LTLC groups (Table 1). In TMT-A, completion time was 76.2 ± 53.5 s and 50.4 ± 21.3 s in the NSTLC and LTLC groups, respectively. Z-Scores were significantly lower in the LTLC group than in the NSTLC group (−0.8 ± 1.2 vs. 2.2 ± 3.6, $p = 0.017$, Figure 2). On TMT-

B, completion times were 138.9 ± 77.7 s and 159.8 ± 107.1 s in the LTLC and NSTLC groups, respectively, whereas one patient in the NSTLC group was unable to complete the TMT-B test. Z-Scores were not significantly different between the LTLC and NSTLC groups (0.3 ± 2.3 vs. 0.4 ± 1.4, Figure 2). In the four subtests of the WMS-R including logical memory I, logical memory II, digit span, and visual memory span, each score and the corresponding Z-score were 18.4 ± 7.5, −0.2 ± 1.0, 12.1 ± 3.3 and −0.1 ± 0.9, and 15.4 ± 2.2, 0 ± 0.7, 14.1 ± 7.7, and −0.1 ± 1.0 points, respectively, and there were no significant differences between the LTLC and NSTLC groups (Figure 2).

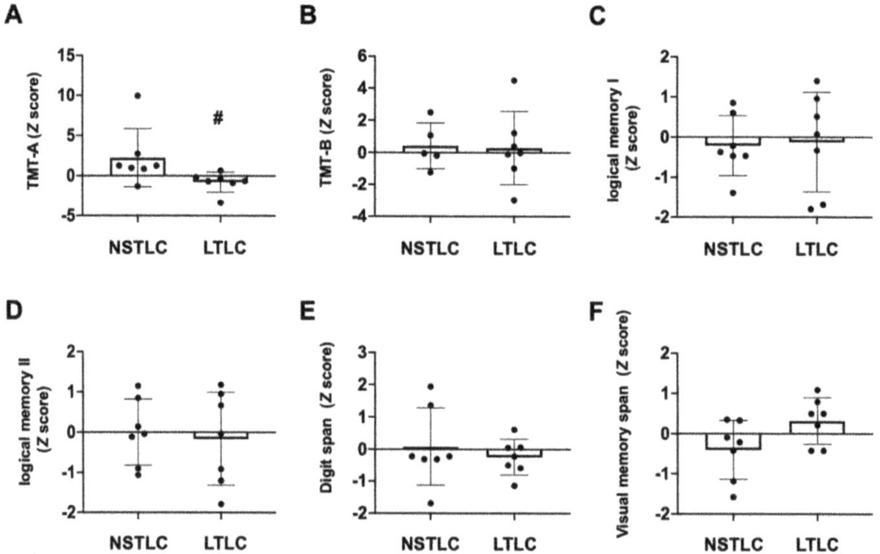

Figure 2. Comparison of neuropsychological tests between patients treated with NSTLC and with LTLC. Dot plots showing Z (standard score) of TMT-A (**A**), TMT-B (**B**), subsets of WMS-R logical memory I (**C**)/II (**D**), digit span (**E**), and visual span (**F**) between hemodialysis patients treated with NSTLC and with LTLC. The Mann–Whitney test was used for comparison. Z (Standard score) = x (raw score) − μ (mean score)/σ (standard deviation); # $p < 0.05$. LTLC, long-term L-carnitine treatment; NSTLC, no or short-term L-carnitine treatment; TMT = trail making test.

3.5. DTI Analysis in Hemodialysis Patients with NSTLC and LTLC

On TBSS, as shown in Figure 3, group comparison showed that there were lower FA and higher AD, RD, and MD in the NSTLC group than in the HC group in many white-matter tracts, whereas there were no significant differences between the LTLC and NSTLC groups and between the HC and LTLC groups (Table 2). TOI analyses showed that FA values in Latr, Ratr, Lcs, Rcs, Lcc, Rcc, fm, fmi, Lifof, Rifof, Lilf, Lslf, Rslf, and Luf were lower in the NSTLC group than in the HC group ($p < 0.05$, Table S2, Supplementary Materials). There were no differences in AD values (Table S3, Supplementary Materials). RD values in Latr, Ratr, Lcs, Rcs, Lcc, Rcc, fm, fmi, Lifof, Rifof, Lilf, Lslf, Rslf, Luf, and Lslftemp, and in fm and Rslf were higher in the NSTLC group and in both the NSTLC and LTLC groups than in the HC group, respectively ($p < 0.05$, Table S4, Supplementary Materials). MD values in Latr, Ratr, fm, fmi, Lifof, Rifof, Lilf, Lslf, Luf, Ruf, Lslftemp, and Rslftemp were higher in the NSTLC group, and MD values in Rslf were higher in the NSTLC and LTLC groups than in the HC group ($p < 0.05$, Table S5, Supplementary Materials).

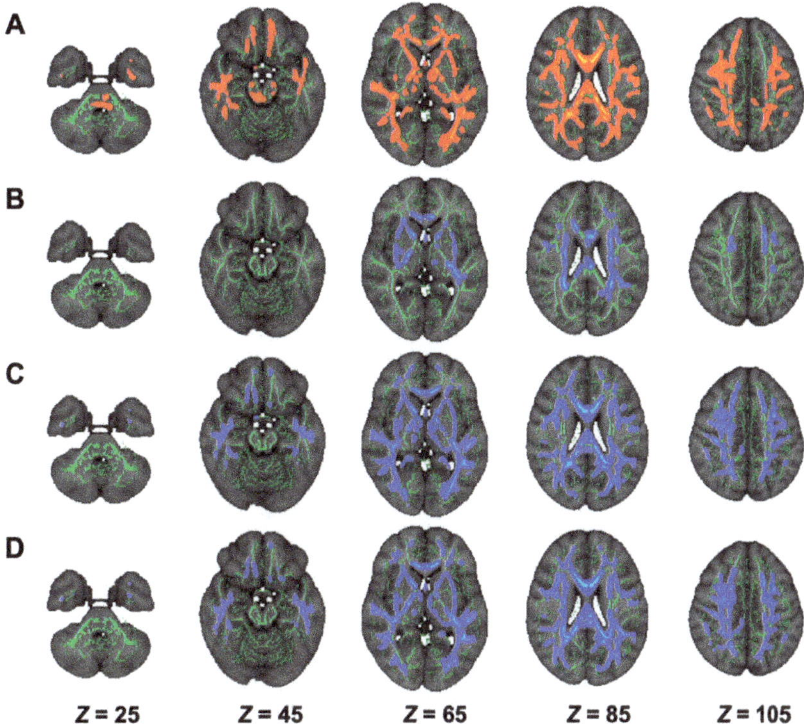

Figure 3. Comparison of TBSS of DTI among hemodialysis patients treated with NSTLC and with LTLC and healthy controls. Comparisons of DTI (FA, (**A**); AD, (**B**); RD, (**C**); and MD, (**D**)) indices among hemodialysis patients treated with NSTLC and LTLC and healthy controls are shown. TBSS analyses show that hemodialysis patients treated with NSTLC have significantly lower FA and significantly higher MD, AD, and RD than healthy controls ($p < 0.05$, (**A–D**)), whereas there are no significant differences between hemodialysis patients treated with NSTLC and LTLC, and hemodialysis patients treated with LTLC and healthy controls (data not shown). The skeleton is presented in green. To aid visualization, the results are thickened using the fill script implemented in the FMRIB Software Library. AD, axial diffusivity; DTI, diffusion tensor imaging; FA, fractional anisotropy; LTLC, long-term L-carnitine treatment; MD, mean diffusivity; NSTLC, no or short-term L-carnitine treatment; RD, radial diffusivity; TBSS, tract-based spatial statistics.

Table 2. Tract-based spatial statistics analysis of diffusion tensor and diffusion kurtosis imaging and neurite orientation dispersion and density imaging in hemodialysis patients and healthy controls.

Modality	Contrast	Cluster Size	Anatomical Region	Peak *t*-Value	Peak MNI Coordinates (X, Y, Z)
			DTI		
FA	HC > NSTLC	48,825	Bilateral ATR, corticospinal tract, CCG, forceps minor and major, IFOF, ILF, SLF, SLF temporal part, medial lemniscus, CP, ALIC, PLIC, retrolenticular part of IC, ACR, SCR, PCR, PTR, SS, external capsule, fornix stria terminalis, SFOF, tapetum; left UF, corticospinal tract, ICP, UF; right SCP; MCP, pontine crossing tract, genu, body and splenium of CC, fornix	6.74	(74, 69, 105)

Table 2. *Cont.*

Modality	Contrast	Cluster Size	Anatomical Region	Peak *t*-Value	Peak MNI Coordinates (*X, Y, Z*)
AD	HC < NSTLC	10,699	Bilateral ATR, corticospinal tract, IFOF, SLF, ALIC, PLIC, retrolenticular part of IC, ACR, SCR, PCR, PTR, external capsule, fornix stria terminalis, SFOF; left ILF, SS, tapetum; forceps minor, UF, genu, body and splenium of CC, fornix	5.85	(50, 125, 100)
RD	HC < NSTLC	49,556	Bilateral ATR, corticospinal tract, CCG, IFOF, ILF, SLF, UF, SLF temporal part, CP, ALIC, PLIC, retrolenticular part of IC, ACR, SCR, PCR, PTR, SS, external capsule, fornix stria terminalis, SFOF, tapetum; forceps minor and major, genu, body and splenium of CC, fornix	7.43	(140, 117, 50)
MD	HC < NSTLC	43,676	Bilateral ATR, corticospinal tract, CCG, IFOF, ILF, SLF, UF, SLF temporal part, ALIC, PLIC, retrolenticular part of IC, ACR, SCR, PCR, PTR, SS, external capsule, fornix stria terminalis, SFOF, tapetum; left CHp; forceps minor and major, genu, body and splenium of CC, fornix	7.10	(53, 104, 105)

DKI					
AK	HC > NSTLC	15,653	Bilateral corticospinal tract, IFOF, ILF, SLF, PLIC, retrolenticular part of IC, ACR, SCR, PCR, PTR, external capsule, fornix stria terminalis, tapetum; right ATR, cingulum hippocampus, UF, SS, CCG, CHp; forceps minor and major, body and splenium of CC	6.03	(53, 65, 61)
	HC > LTLC	16,138	Bilateral corticospinal tract, CHp, IFOF, ILF, SLF, SLF temporal part, corticospinal tract, medial lemniscus, SCP, CP, PLIC, retrolenticular part of IC, SCR, PCR, PTR, SS, fornix stria terminalis, tapetum; left ATR, UF, ALIC, ACR, external capsule, SFOF; forceps major, MCP, pontine crossing tract, body and splenium of CC	6.19	(55, 92, 82)
RK	HC > NSTLC	45,239	Bilateral ATR, corticospinal tract, CCG, IFOF, ILF, SLF, UF, SLF temporal part, ALIC, retrolenticular part of IC, ACR, SCR, PCR, PTR, SS, external capsule, fornix stria terminalis, SFOF; right UF, tapetum; forceps minor and major, genu, body and splenium of CC	6.11	(98, 172, 110)
	HC > LTLC	69	forceps minor	4.89	(79, 177, 101)
MK	HC > NSTLC	28,546	Bilateral ATR, IFOF, ILF, SLF, ALIC, ACR, SCR, PCR, PTR, SS, external capsule, SFOF; left corticospinal tract, UF, retrolenticular part of IC; forceps minor and major, genu, body and splenium of CC	6.82	(97, 141, 131)

NODDI					
ICVF	HC > NSTLC	67,959	Bilateral ATR, corticospinal tract, CCG, IFOF, ILF, SLF, UF, SLF temporal part, CP, ALIC, PLIC, retrolenticular part of IC, ACR, SCR, PCR, PTR, SS, external capsule, CHp, fornix stria terminalis, SFOF, tapetum; left UF; right CHp; forceps minor and major, genu, body and splenium of CC	8.15	(106, 82, 84)

Table 2. *Cont.*

Modality	Contrast	Cluster Size	Anatomical Region	Peak *t*-Value	Peak MNI Coordinates (*X, Y, Z*)
	HC > LTLC	38,021	Bilateral ATR, corticospinal tract, CCG, IFOF, ILF, SLF, UF, SLF temporal part, ALIC, PLIC, retrolenticular part of IC, ACR, SCR, PCR, PTR, SS, external capsule, CHp, fornix stria terminalis, SFOF; left tapetum; right CHp, CP; forceps minor and major, genu, body and splenium of CC	6.22	(113, 63, 103)
ISO	HC > LTLC	3978	right ATR, Inferior fronto-occipital fasciculus, ILF, UF, retrolenticular part of IC, ACR, SCR, PTR, SS; forceps minor and major	5.88	(48, 116, 49)
	NSTLC > LTLC	8312	Bilateral corticospinal tract, IFOF, ILF, SLF, UF, SCP, CP, retrolenticular part of IC, PTR, SS; right ATR, CHp, SLF temporal part, ALIC, PLIC, external capsule, fornix stria terminalis; forceps minor and major, MCP, pontine crossing tract	7.52	(129, 83, 68)

DTI = diffusion tensor imaging; DKI = diffusion kurtosis imaging; NODDI = neurite orientation dispersion and density imaging; ICVF = intracellular volume fraction; ODI = orientation dispersion index; ISO = isotropic volume fraction; HC = healthy control; NSTLC = no or short-term L-carnitine treatment; LTLC = long-term L-carnitine treatment; ATR = anterior thalamic radiation; CCG = cingulum in cingulate gyrus; IFOF = inferior fronto-occipital fasciculus; ILF = inferior longitudinal fasciculus; SLF = superior longitudinal fasciculus; CP = cerebral peduncle; ALIC = anterior limb of internal capsule; PLIC = posterior limb of internal capsule; ACR = anterior corona radiata; SCR = superior corona radiata; PCR = posterior corona radiata; PTR = posterior thalamic radiation; SS = striatum; SFOF = superior frontal occipital fasciculus; UF = uncinate fasciculus; ICP = inferior cerebellar peduncle; SCP = superior cerebellar peduncle; MCP = middle cerebellar peduncle; CC = corpus callosum.

3.6. DKI Analysis in Hemodialysis Patients with NSTLC and LTLC

On TBSS, AK was lower in the LTLC and NSTLC groups than in the HC group (Figure 4A,B, Table 2). In RK and MK, bilateral white-matter tracts were widely lower in the NSTLC group than in the HC group, whereas RK in a confined tract was lower in the LTLC group than in the HC group, and there was no difference in MK between the LTLC and HC groups (Figure 4C–E, Table 2). On TOI analyses, RK values in Lcc and Rifof and in Ratr, Rcc, and Lslf were lower in the NSTLC group and in both the NSTLC and LTLC groups than in the HC group ($p < 0.05$), respectively, whereas there were no differences in AK and MK values (Tables S6–S8, Supplementary Materials).

3.7. NODDI Analysis in Hemodialysis Patients with NSTLC and LTLC

On TBSS analysis, the NSTLC and LTLC groups showed significantly lower ICVF in bilateral subcortical white matter tracts than the HC group (Figure 5A,B, Table 2). In TOI analyses, ICVF values in Latr, Ratr, Lcs, Rcs, Lcc, Rcc, fm, fmi, Lifof, Rifof, Lilf, Rilf, Lslf, Rslf, Lslftemp, and Rslftemp were lower in both or either of the NSTLC and LTLC groups than in the HC group ($p < 0.05$, Table S9, Supplementary Materials). In ODI analyses, there were no significant differences among the three groups on TBSS and TOI analyses (Table S10, Supplementary Materials). In ISO analyses, the LTLC group showed lower values in right temporal, frontal, and occipital subcortical tracts than the HC and NSTLC groups (Figure 5C,D, Table 2), whereas there were no differences in TOI values (Table S11, Supplementary Materials).

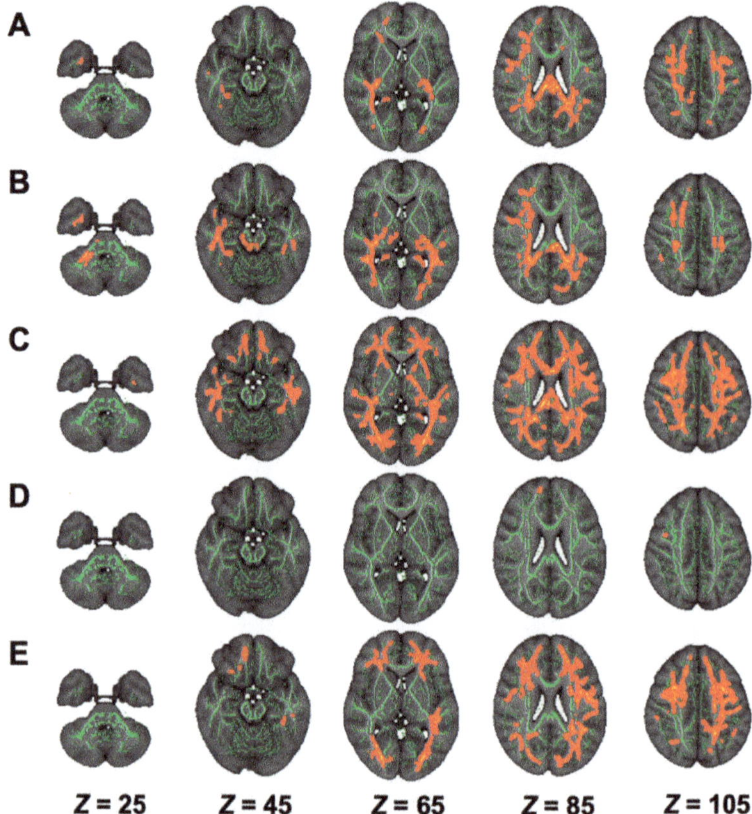

Figure 4. Comparison of TBSS of DKI among hemodialysis patients with NSTLC and LTLC and healthy controls. Comparisons of DKI (AK, (**A**,**B**); RK, (**C**,**D**); and MK, (**E**) indices among hemodialysis patients with NSTLC and LTLC and healthy controls are shown. TBSS analyses show that hemodialysis patients treated with NSTLC have significantly lower AK, RK, and MK than healthy controls ($p < 0.05$, (**A**,**C**,**E**)). Significant but limited low fiber tracts in AK and RK in hemodialysis patients treated with LTLC compared to healthy controls are found ($p < 0.05$, (**B**,**D**)). The skeleton is presented in green. To aid visualization, the results are thickened using the fill script implemented in the FMRIB Software Library. AK, axial kurtosis; DKI, diffusion kurtosis imaging; LTLC, long-term L-carnitine treatment; MK, mean kurtosis; NSTLC, no or short-term L-carnitine treatment; RK, radial kurtosis; TBSS, tract-based spatial statistics.

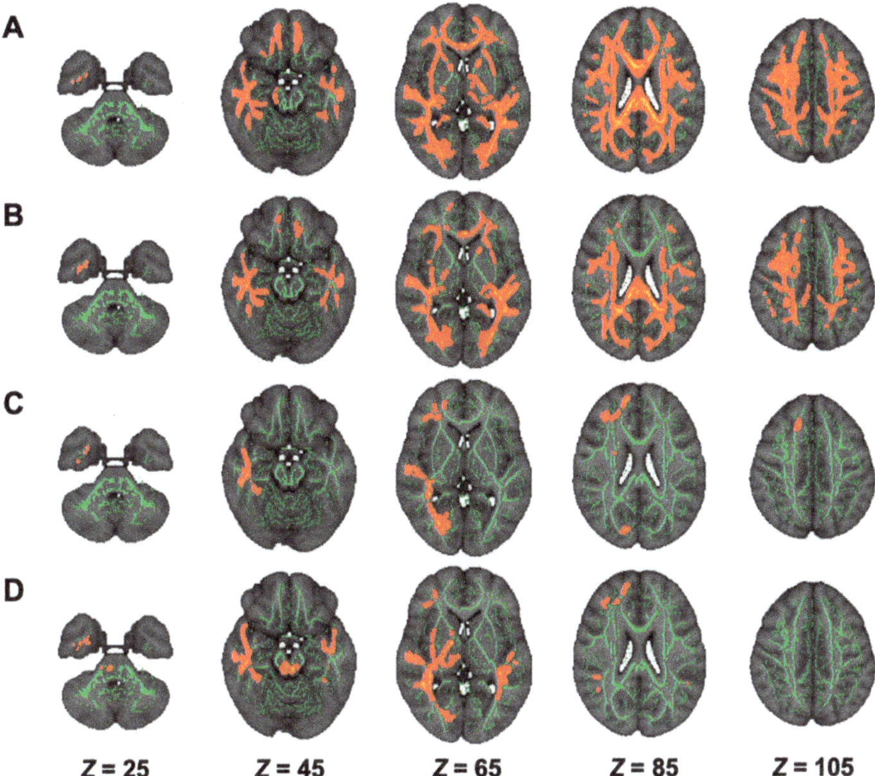

Figure 5. Comparison of TBSS of NODDI among hemodialysis patients treated with NSTLC and LTLC and healthy controls. Comparisons of ICVF (**A**,**B**) and ISO (**C**,**D**) indices among hemodialysis patients treated with NSTLC and LTLC and healthy controls are shown. TBSS analyses show that hemodialysis patients treated with NSTLC (**A**) and LTLC (**B**) have significantly lower ICVF than healthy controls ($p < 0.05$). TBSS analyses show that hemodialysis patients treated with LTLC have significantly lower ISO than hemodialysis patients treated with NSTLC (**C**) and healthy controls (**D**) ($p < 0.05$). The skeleton is presented in green. To aid visualization, the results are thickened using the fill script implemented in the FMRIB Software Library. ICVF, intracellular volume fraction; ISO, isotropic volume fraction; LTLC, long-term L-carnitine treatment; NODDI, neurite orientation dispersion and density imaging; NSTLC, no or short-term L-carnitine treatment; TBSS, tract-based spatial statistics.

3.8. Association of Cognitive Function Tests with dMRI in the Comparison between NSTLC and LTLC Groups

Among the types of cognitive function tests, the LTLC group showed better achievement of TMT-A than the NSTLC group. Significant correlations were shown for the associations of completion times of TMT-A with Ratr ($r = 0.79$, $p = 0.001$) in AD, Lcc ($r = -0.75$, $p = 0.002$), Latr ($r = -0.77$, $p = 0.001$), and Rslf ($r = -0.76$, $p = 0.002$) in AK, and fm ($r = -0.79$, $p = 0.001$) in ICVF (Table 3).

Table 3. Association of completion time of Trail-making test A with tracts-of-interest on DTI, DKI, and NODDI between NSTLC and LTLC groups.

Region	FA		AD		RD		MD		AK		RK		MK		ICVF		ISO	
	R	p	R	p	R	p	R	p	R	p	R	p	R	p	R	p	R	p
Whole			0.628	0.016			0.604	0.022	−0.741	0.002 *	−0.591	0.026	−0.609	0.021	−0.547	0.043		
Lcc	−0.543	0.045							−0.745	0.002 *								
Rcc	−0.609	0.021			0.669	0.009	0.591	0.026			−0.565	0.035						
Latr			0.679	0.008			0.6	0.023	−0.771	0.001 *					−0.613	0.02		
Ratr			0.793	0.001 *	0.554	0.04	0.648	0.012	−0.248	0.392							0.609	0.021
Lcs			0.644	0.013														
Rcs			0.569	0.034			0.559	0.038										
fm					0.569	0.034	0.678	0.008							−0.789	0.001 *		
fmi									−0.574	0.032			−0.644	0.013	−0.534	0.049		
Lifof									−0.613	0.02					−0.582	0.029		
Rifof											−0.648	0.012	−0.591	0.026				
Lslf									−0.613	0.02								
Rslf									−0.758	0.002 *					−0.574	0.032		
Lilf									−0.556	0.039								
Rilf											−0.578	0.03						
Rslftemp											−0.587	0.027						

Spearman's rank correlation coefficient was used for comparison. * Values differed significantly ($p < 0.05$, FDR corrected). LCAR = L-carnitine; LTLC = large amount of total L-carnitine treatment; NSTLC = no or short amount of total L-carnitine treatment; DTI = diffusion tensor imaging; DKI = diffusion kurtosis imaging; NODDI = neurite orientation dispersion and density imaging; FA = fractional anisotropy; AD = axial diffusivity; RD = radial diffusivity; MD = mean diffusivity; AK = axial kurtosis; RK = radial kurtosis; MK = mean kurtosis; ICVF = intra-cellular volume fraction; ISO = isotropic volume fraction; ODI = orientation dispersion index; FDR = false detection rate; L(R)atr = left(right) anterior thalamic radiation; L(R)cs = corticospinal tract; L(R)cc = cingulum (cingulate gyrus); L(R)ch = cingulum (hippocampus); fm = forceps major; fmi = forceps minor; L(R)ifof = left(right) inferior fronto-occipital fasciculus; L(R)ilf = left(right) inferior longitudinal fasciculus; L(R)slf = left(right) superior longitudinal fasciculus; L(R)uf = left(right) uncinate fasciculus; L(R)slftemp = left(right) superior longitudinal fasciculus temporal part.

3.9. Histopathological Characteristics of an Autopsy Brain

A 75 years old Japanese man with NSTLC, with a previous history of diabetes mellitus, diabetic retinopathy, and nephropathy, developed intracranial hemorrhage in the right thalamus and was admitted to our hospital. Despite acute therapy for thalamic hemorrhage, such as intravenous blood pressure-lowering drugs, glycerin, and hemostatic agents, the thalamic hemorrhage worsened, and he died 9 days after admission. Postmortem examination showed that right thalamic hemorrhage was comorbid with intraventricular rupture, but cerebral white matter was preserved depending on the location where microscopic analyses were carried out (Figure 6A,B). On KB staining, moderate rarefaction of white matter together with loss of myelin sheaths was found in the subcortical white matter of the superior frontal gyrus (Figure 6C,D). The NFH staining showed that axons were less sparse in the superior frontal gyrus (Figure 6E). Phosphorylated tau inclusions and accumulations of amyloid beta in the hippocampus were not apparent (Figure 6F,G).

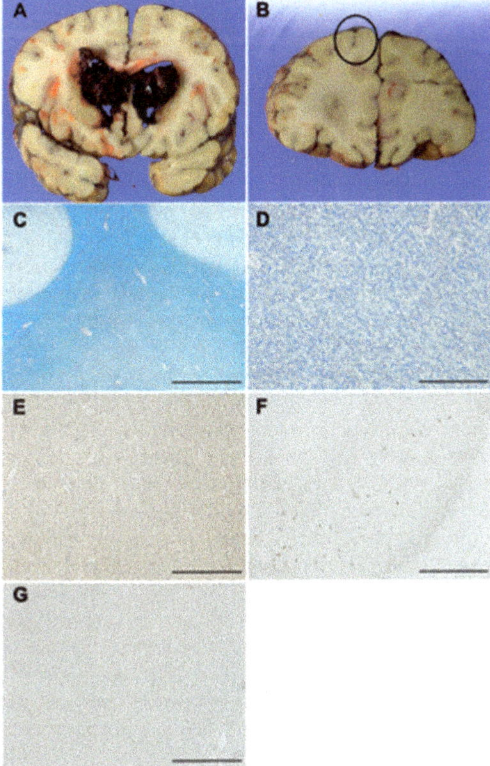

Figure 6. Postmortem study of a case of NSTLC. Macroscopic view of the bilateral cerebral hemispheres shows that right thalamic hemorrhage is comorbid with intraventricular rupture (**A**), but subcortical white mater is preserved (**B**), where histological analysis was carried out ((**B**), circle; (**C–E**)). Photomicrographs of Klüber–Barrera ((**C**), scale bar: 5 mm; (**D**), scale bar: 200 μm), neurofilament heavy weight ((**E**), scale bar: 200 μm), phosphorylated tau (**F**), scale bar: 1 mm), and amyloid beta ((**G**), scale bar: 1 mm) staining.

4. Discussion

4.1. Main Findings

The present study first explored the disorder of white-matter tracts using dMRI and the impact of L-carnitine treatment for the white-matter tract injuries, as well as cognitive dysfunction, in hemodialysis patients. The principal findings of the current study were that impairment of white-matter tracts was robustly detected in hemodialysis patients with pathological confirmation in an autopsy case, and these injuries were alleviated by treatment with LTLC, along with reduction of hs-CRP levels. Furthermore, hemodialysis patients treated with LTLC displayed better performance on TMT-A than hemodialysis patients with NSTLC, and specific white-matter tracts that contributed to the achievement of TMA-A were identified.

Previous studies showed that hemodialysis patients had mean MMSE scores ranging from 20.7 to 25.7 points, and 24–87% of hemodialysis populations were categorized as having dementia [2–4,25–28]. In the current study, MMSE and MoCA-J scores were 25.4 ± 3.5 and 23.9 ± 3.2 points, respectively, and the majority of the cases were considered to have mild cognitive impairment, because the cutoff values of MMSE and MoCA-J were found to be 27 and 25 points, respectively [29–31]. Importantly, the current data showed that low FA, high RD and MD, and low RK were clearly seen in hemodialysis patients, particularly in those treated with NSTLC. Furthermore, an autopsy study showed rarefaction of white matter and loss of myelin sheaths without the pathological hallmarks of Alzheimer's disease, which were consistent with those of vascular dementia [12,13]. Thus, the present data showed that damage to microstructural organizations in cerebral white matter can be a fundamental cause of cognitive decline in hemodialysis patients, which was clearly shown by the correlation between dMRI and histopathology.

Several mechanisms for the development of cognitive dysfunction related to white-matter diseases in hemodialysis patients have been proposed. First, concomitant atherosclerotic vascular risk factors contributed to the cerebral small vessel diseases yielding silent infarcts, white-matter lesions, and microbleeds, thereby causing vascular dementia [32–35]. Second, hemodialysis itself altered cerebral blood flow, which was previously shown by positron emission tomography and transcranial Doppler [36,37]. Induction of acute fluid shifts, intravascular volume loss, and brain edema, thereby leading to disorders of cerebral tissue and metabolism, could occur. Third, oxidative stress, chronic inflammation, and coagulopathy that were increased in end-stage renal disease may induce endothelial injury and neuronal damage [3]. Alternatively, uremic toxins such as hyperhomocysteinemia, guanidine compound, and cystatin-C have a role in neurodegeneration [38]. The present data showed that long-term L-carnitine therapy improved white-matter structural damage and TMT-A performance, along with a reduction in serum hs-CRP levels. It was shown that L-carnitine reduced markers of oxidative stress and inflammation in end-stage renal disease [14]. Our previous experimental research demonstrated that continuous supplementation of L-carnitine for 28 days after cerebral chronic hypoperfusion resulted in suppression of oxidative stress in oligodendrocytes, facilitation of myelin sheaths and axonal outgrowth via the phosphatase and tensin homologue deleted on chromosome ten/Akt/mammalian target of rapamycin pathway, protection from mitochondrial dysfunction, and improvement of cognitive dysfunction. Generally, 80% of the total plasma carnitine is free carnitine, and 20% is acylated carnitine, with a normal acyl/free carnitine ratio of 0.25 [39]. Free carnitine not only transports long-chain fatty acids to mitochondria for subsequent β-oxidation, but also acts as a scavenger by eliminating toxic acyl compounds from mitochondria; thus, an acyl/free carnitine ratio >0.4 indicates carnitine insufficiency [39]. In the present data, although the acyl/free carnitine ratio was still >0.4 after L-carnitine therapy, substantial increases in free carnitine levels could robustly increase β-oxidation and ATP synthesis, thereby exerting pleiotropic effects and protecting white-matter microstructure in hemodialysis patients. Furthermore, the present data suggest that L-carnitine could be a potential candidate for novel therapeutic agents for vascular cognitive dysfunction other than in hemodialysis patients.

In the present comprehensive dMRI data, white-matter microstructural damage showing low FA, high RD and MD, and low RK on DTI and DKI was more critical in NSTLC. On the other hand, NODDI showed comparably low ICVF in LTLC and NSTLC, whereas relatively lower ISO in LTLC comparing with NSTLC was seen in limited tracts. ICVF indicates the density of neurites including axons and dendrites based on intracellular diffusion, and ISO indicates the volume fraction of isotropic diffusion [7]. Considering the previous experimental and the current results, although considerably lower neurite density due to damage of neural tissues could occur, long-term treatment with L-carnitine may alleviate injuries of white-matter microstructures, possibly due to suppression of neuroinflammation, in hemodialysis patients. However, the sample size in the current study was small, and a future prospective study with a larger sample size is warranted.

The LTLC group showed better achievement on the TMT-A than hemodialysis patients in the NSTLC groups. TMT is a task to evaluate the ability to flexibly switch attention between competing task-set representations and is denoted by the total time to completion [40]. The TMT-A is considered to reflect a baseline measure of psychomotor speed, visuospatial search, and target-directed motor tracking [40,41]. Fundamentally, multiple cognitive processes are thought to be implicated in performing the TMT; therefore, different brain regions and connecting fiber tracts are associated with completing the TMT. Lesion–symptom mapping studies showed that the right frontal area, left rostral anterior cingulate cortex, and left dorsomedial prefrontal lobe were related to impairments of TMT performance in patients with central nervous system disorders such as stroke and brain tumors [42–44]. The present comprehensive dMRI data showed that completion times for TMT-A correlated with disorders of bilateral anterior thalamic radiations, the left cingulum in the cingulate gyrus, the right superior longitudinal fasciculus, and forceps major. Thus, the combination of DTI, DKI, and NODDI elucidated the neuroanatomical connections related to performance on the TMT-A in hemodialysis patients for the first time.

4.2. Limitations

Some potential limitations of the current study must be considered when interpreting the present results. First, the sample size was small, and this study was a nonrandomized, cross-sectional study analyzing the therapeutic effects of L-carnitine for cognitive decline and microstructural damage on dMRI in hemodialysis patients. Although there was a significant difference in the total amount of L-carnitine between the LTLC and NSTLC groups, the protocol of L-carnitine treatment was quite complicated, and the NSTLC group was heterogenous in that it consisted of patients without L-carnitine treatment and patients with short-term L-carnitine treatment. Thus, to explore the protective effects of L-carnitine for disorders of white-matter fiber tracts and cognitive function, large-scale, prospective longitudinal studies with an appropriate trial design are warranted. Furthermore, it is important to explore the association of serum carnitine concentrations with microstructural damage of cerebral white matter. Second, only one postmortem examination was performed in the current study. The types of end-stage renal diseases in the enrolled patients varied; therefore, there may have been heterogeneity in postmortem brain pathology. Third, other medications including statins, antiplatelet drugs, and angiotensin II receptor blockers might have antiatherogenic effects and could have affected the microstructural damage of white-matter tracts and the neuropsychological status in hemodialysis patients.

5. Conclusions

The current data showed that L-carnitine may have neuroprotective roles against the white-matter microstructural damage and cognitive impairment in hemodialysis patients. It was found that DTI, DKI, and NODDI could certainly detect abnormalities of white-matter fiber tracts in hemodialysis patients, consistent with rarefaction of white matter and loss of myelin sheaths on histopathology. Long-term treatment with L-carnitine may ameliorate white-matter microstructural injuries by possibly suppressing neuroinflammation and improve achievement of TMT-A performance linked with protecting some candidate fiber tracts. Long-term treatment with L-carnitine can be a novel therapeutic approach for vascular cognitive dementia in hemodialysis patients.

Supplementary Materials: The following are available online at https://www.mdpi.com/article/10.3390/nu13041292/s1: Table S1. Means and standard deviations of scores on neuropsychological tests in Japanese subjects by age group; Table S2. Tract-of-interest analysis for fractional anisotropy among hemodialysis patients with no or short-term L-carnitine treatment, long-term L-carnitine treatment, and healthy controls; Table S3. Tract-of-interest analysis for axial diffusivity among hemodialysis patients with no or short-term L-carnitine treatment, long-term L-carnitine treatment, and healthy controls; Table S4. Tract-of-interest analysis for radial diffusivity among hemodialysis patients with no or short-term L-carnitine treatment, long-term L-carnitine treatment, and healthy controls; Table S5. Tract-of-interest analysis for mean diffusivity among hemodialysis patients with no or short-term L-carnitine treatment, long-term L-carnitine treatment, and healthy controls; Table S6. Tract-of-interest analysis for axial kurtosis among hemodialysis patients with no or short-term L-carnitine treatment, long-term L-carnitine treatment, and healthy controls; Table S7. Tract-of-interest analysis for radial kurtosis among hemodialysis patients with no or short-term L-carnitine treatment, long-term L-carnitine treatment, and healthy controls; Table S8. Tract-of interest analysis for mean kurtosis among hemodialysis patients with no or short-term L-carnitine treatment, long-term L-carnitine treatment, and healthy controls; Table S9. Tract-of-interest analysis for intra-cellular volume fraction among hemodialysis patients with no or short-term L-carnitine treatment, long-term L-carnitine treatment, and healthy controls; Table S10. Tract-of-interest analysis for the orientation dispersion index among hemodialysis patients with no or short-term L-carnitine treatment, long-term L-carnitine treatment, and healthy controls; Table S11. Region-of-interest analysis for isotropic volume fraction among hemodialysis patients with no or short-term L-carnitine treatment, long-term L-carnitine treatment, and healthy controls.

Author Contributions: Conceptualization, Y.U., A.S. (Asami Saito), J.N., H.I., and S.A.; methodology, Y.U., A.S. (Asami Saito), K.K., and S.A.; investigation, Y.U., A.S. (Asami Saito), J.N., K.K., D.T., Y.M., H.I., C.A., A.S. (Atsuhiko Shindo), K.S., N.M., and K.Y.; formal analysis, Y.U., A.S. (Asami Saito), J.N., K.K., D.T., Y.M., H.I., C.A., A.S. (Atsuhiko Shindo), K.S., N.M., and K.Y.; resources, Y.U. and N.H.; writing—original draft, Y.U., A.S. (Asami Saito), K.K., and C.A.; writing—review and editing, all authors; supervision, T.U., Y.S., S.A., and N.H. All authors have read and agreed to the published version of the manuscript.

Funding: This work was funded, in part, by the Japan Society for the Promotion of Science (grant numbers 17K09764 and 20K07909 to Ueno), a Grant-in-Aid (S1411006) from the Foundation of Strategic Research Projects in Private Universities from the Ministry of Education, Culture, Sports, Science, and Technology, a grant from the Takeda Science Foundation, and a grant from the Kenzo Suzuki Foundation.

Institutional Review Board Statement: This study was conducted in accordance with the Declaration of Helsinki. The independent ethics committee of Juntendo University Hospital approved this study (16-170).

Informed Consent Statement: All study subjects were given an explanation of the study, and written, informed consent was obtained for the study objective, MRI and cognitive function tests, enrolment in the study, autopsy study, and ensuring confidentiality of information.

Data Availability Statement: The datasets used and analyzed during the current study are available from the corresponding author upon reasonable request.

Conflicts of Interest: The authors declare no conflict of interest.

References

1. Thomas, B.; Wulf, S.; Bikbov, B.; Perico, N.; Cortinovis, M.; de Vaccaro, K.C.; Flaxman, A.; Peterson, H.; Delossantos, A.; Haring, D.; et al. Maintenance Dialysis throughout the World in Years 1990 and 2010. *J. Am. Soc. Nephrol.* **2015**, *26*, 2621–2633. [CrossRef] [PubMed]
2. Kurella, M.; Mapes, D.L.; Port, F.K.; Chertow, G.M. Correlates and outcomes of dementia among dialysis patients: The Dialysis Outcomes and Practice Patterns Study. *Nephrol. Dial. Transplant.* **2006**, *21*, 2543–2548. [CrossRef] [PubMed]
3. Bugnicourt, J.M.; Godefroy, O.; Chillon, J.M.; Choukroun, G.; Massy, Z.A. Cognitive disorders and dementia in CKD: The neglected kidney-brain axis. *J. Am. Soc. Nephrol.* **2013**, *24*, 353–363. [CrossRef]

4. Murray, A.M.; Tupper, D.E.; Knopman, D.S.; Gilbertson, D.T.; Pederson, S.L.; Li, S.; Smith, G.E.; Hochhalter, A.K.; Collins, A.J.; Kane, R.L. Cognitive impairment in hemodialysis patients is common. *Neurology* **2006**, *67*, 216–223. [CrossRef] [PubMed]

5. Assaf, Y.; Pasternak, O. Diffusion tensor imaging (DTI)-based white matter mapping in brain research: A review. *J. Mol. Neurosci.* **2008**, *34*, 51–61. [CrossRef]

6. Jensen, J.H.; Helpern, J.A.; Ramani, A.; Lu, H.; Kaczynski, K. Diffusional kurtosis imaging: The quantification of non-gaussian water diffusion by means of magnetic resonance imaging. *Magn. Reson. Med.* **2005**, *53*, 1432–1440. [CrossRef] [PubMed]

7. Kamagata, K.; Zalesky, A.; Hatano, T.; Ueda, R.; Di Biase, M.A.; Okuzumi, A.; Shimoji, K.; Hori, M.; Caeyenberghs, K.; Pantelis, C.; et al. Gray Matter Abnormalities in Idiopathic Parkinson's Disease: Evaluation by Diffusional Kurtosis Imaging and Neurite Orientation Dispersion and Density Imaging. *Hum. Brain Mapp.* **2017**. [CrossRef]

8. Maillard, P.; Fletcher, E.; Singh, B.; Martinez, O.; Johnson, D.K.; Olichney, J.M.; Farias, S.T.; DeCarli, C. Cerebral white matter free water: A sensitive biomarker of cognition and function. *Neurology* **2019**, *92*, e2221–e2231. [CrossRef]

9. Williams, O.A.; Zeestraten, E.A.; Benjamin, P.; Lambert, C.; Lawrence, A.J.; Mackinnon, A.D.; Morris, R.G.; Markus, H.S.; Barrick, T.R.; Charlton, R.A. Predicting Dementia in Cerebral Small Vessel Disease Using an Automatic Diffusion Tensor Image Segmentation Technique. *Stroke* **2019**, *50*, 2775–2782. [CrossRef]

10. Drew, D.A.; Koo, B.B.; Bhadelia, R.; Weiner, D.E.; Duncan, S.; la Garza, M.M.; Gupta, A.; Tighiouart, H.; Scott, T.; Sarnak, M.J. White matter damage in maintenance hemodialysis patients: A diffusion tensor imaging study. *BMC Nephrol.* **2017**, *18*, 213. [CrossRef]

11. Chou, M.C.; Hsieh, T.J.; Lin, Y.L.; Hsieh, Y.T.; Li, W.Z.; Chang, J.M.; Ko, C.H.; Kao, E.F.; Jaw, T.S.; Liu, G.C. Widespread white matter alterations in patients with end-stage renal disease: A voxelwise diffusion tensor imaging study. *AJNR Am. J. Neuroradiol.* **2013**, *34*, 1945–1951. [CrossRef]

12. Ueno, Y.; Zhang, N.; Miyamoto, N.; Tanaka, R.; Hattori, N.; Urabe, T. Edaravone attenuates white matter lesions through endothelial protection in a rat chronic hypoperfusion model. *Neuroscience* **2009**, *162*, 317–327. [CrossRef]

13. Ueno, Y.; Koike, M.; Shimada, Y.; Shimura, H.; Hira, K.; Tanaka, R.; Uchiyama, Y.; Hattori, N.; Urabe, T. L-carnitine enhances axonal plasticity and improves white-matter lesions after chronic hypoperfusion in rat brain. *J. Cereb. Blood Flow Metab.* **2015**, *35*, 382–391. [CrossRef]

14. Ringseis, R.; Keller, J.; Eder, K. Mechanisms underlying the anti-wasting effect of L-carnitine supplementation under pathologic conditions: Evidence from experimental and clinical studies. *Eur. J. Nutr.* **2013**, *52*, 1421–1442. [CrossRef] [PubMed]

15. Higuchi, T.; Abe, M.; Yamazaki, T.; Okawa, E.; Ando, H.; Hotta, S.; Oikawa, O.; Kikuchi, F.; Okada, K.; Soma, M. Levocarnitine Improves Cardiac Function in Hemodialysis Patients With Left Ventricular Hypertrophy: A Randomized Controlled Trial. *Am. J. Kidney Dis.* **2016**, *67*, 260–270. [CrossRef]

16. Delaney, C.L.; Spark, J.I.; Thomas, J.; Wong, Y.T.; Chan, L.T.; Miller, M.D. A systematic review to evaluate the effectiveness of carnitine supplementation in improving walking performance among individuals with intermittent claudication. *Atherosclerosis* **2013**, *229*, 1–9. [CrossRef]

17. Mastropietro, A.; Rizzo, G.; Fontana, L.; Figini, M.; Bernardini, B.; Straffi, L.; Marcheselli, S.; Ghirmai, S.; Nuzzi, N.P.; Malosio, M.L.; et al. Microstructural characterization of corticospinal tract in subacute and chronic stroke patients with distal lesions by means of advanced diffusion MRI. *Neuroradiology* **2019**, *61*, 1033–1045. [CrossRef] [PubMed]

18. Nemanich, S.T.; Mueller, B.A.; Gillick, B.T. Neurite orientation dispersion and density imaging quantifies corticospinal tract microstructural organization in children with unilateral cerebral palsy. *Hum. Brain Mapp.* **2019**, *40*, 4888–4900. [CrossRef] [PubMed]

19. Shimada, Y.; Ueno, Y.; Tanaka, Y.; Okuzumi, A.; Miyamoto, N.; Yamashiro, K.; Tanaka, R.; Hattori, N.; Urabe, T. Aging, aortic arch calcification, and multiple brain infarcts are associated with aortogenic brain embolism. *Cerebrovasc. Dis.* **2013**, *35*, 282–290. [CrossRef] [PubMed]

20. Abe, M.; Suzuki, K.; Okada, K.; Miura, R.; Fujii, T.; Etsurou, M.; Yamadori, A. Normative data on tests for frontal lobe functions: Trail Making Test, Verbal fluency, Wisconsin Card Sorting Test (Keio version). *No To Shinkei* **2004**, *56*, 567–574.

21. Fazekas, F.; Chawluk, J.B.; Alavi, A.; Hurtig, H.I.; Zimmerman, R.A. MR signal abnormalities at 1.5 T in Alzheimer's dementia and normal aging. *AJR Am. J. Roentgenol.* **1987**, *149*, 351–356. [CrossRef] [PubMed]

22. Andersson, J.L.R.; Sotiropoulos, S.N. An integrated approach to correction for off-resonance effects and subject movement in diffusion MR imaging. *Neuroimage* **2016**, *125*, 1063–1078. [CrossRef]

23. Smith, S.M.; Jenkinson, M.; Johansen-Berg, H.; Rueckert, D.; Nichols, T.E.; Mackay, C.E.; Watkins, K.E.; Ciccarelli, O.; Cader, M.Z.; Matthews, P.M.; et al. Tract-based spatial statistics: Voxelwise analysis of multi-subject diffusion data. *Neuroimage* **2006**, *31*, 1487–1505. [CrossRef]

24. Cohen, J. A power primer. *Psychol. Bull.* **1992**, *112*, 155–159. [CrossRef] [PubMed]

25. Fazekas, G.; Fazekas, F.; Schmidt, R.; Kapeller, P.; Offenbacher, H.; Krejs, G.J. Brain MRI findings and cognitive impairment in patients undergoing chronic hemodialysis treatment. *J. Neurol. Sci.* **1995**, *134*, 83–88. [CrossRef]

26. Kurella, M.; Chertow, G.M.; Luan, J.; Yaffe, K. Cognitive impairment in chronic kidney disease. *J. Am. Geriatr. Soc.* **2004**, *52*, 1863–1869. [CrossRef]

27. Sehgal, A.R.; Grey, S.F.; DeOreo, P.B.; Whitehouse, P.J. Prevalence, recognition, and implications of mental impairment among hemodialysis patients. *Am. J. Kidney Dis.* **1997**, *30*, 41–49. [CrossRef]

28. Antoine, V.; Souid, M.; Andre, C.; Barthelemy, F.; Saint-Jean, O. Symptoms and quality of life of hemodialysis patients aged 75 and over. *Nephrologie* **2004**, *25*, 89–96. [PubMed]
29. Ng, T.P.; Feng, L.; Nyunt, M.S.; Feng, L.; Gao, Q.; Lim, M.L.; Collinson, S.L.; Chong, M.S.; Lim, W.S.; Lee, T.S.; et al. Metabolic Syndrome and the Risk of Mild Cognitive Impairment and Progression to Dementia: Follow-up of the Singapore Longitudinal Ageing Study Cohort. *JAMA Neurol.* **2016**, *73*, 456–463. [CrossRef] [PubMed]
30. O'Bryant, S.E.; Humphreys, J.D.; Smith, G.E.; Ivnik, R.J.; Graff-Radford, N.R.; Petersen, R.C.; Lucas, J.A. Detecting dementia with the mini-mental state examination in highly educated individuals. *Arch. Neurol.* **2008**, *65*, 963–967. [CrossRef] [PubMed]
31. Fujiwara, Y.; Suzuki, H.; Yasunaga, M.; Sugiyama, M.; Ijuin, M.; Sakuma, N.; Inagaki, H.; Iwasa, H.; Ura, C.; Yatomi, N.; et al. Brief screening tool for mild cognitive impairment in older Japanese: Validation of the Japanese version of the Montreal Cognitive Assessment. *Geriatr. Gerontol. Int.* **2010**, *10*, 225–232. [CrossRef]
32. Snowdon, D.A.; Greiner, L.H.; Mortimer, J.A.; Riley, K.P.; Greiner, P.A.; Markesbery, W.R. Brain infarction and the clinical expression of Alzheimer disease. The Nun Study. *JAMA* **1997**, *277*, 813–817. [CrossRef]
33. O'Brien, J.T.; Erkinjuntti, T.; Reisberg, B.; Roman, G.; Sawada, T.; Pantoni, L.; Bowler, J.V.; Ballard, C.; DeCarli, C.; Gorelick, P.B.; et al. Vascular cognitive impairment. *Lancet Neurol.* **2003**, *2*, 89–98. [CrossRef]
34. Yakushiji, Y.; Nishiyama, M.; Yakushiji, S.; Hirotsu, T.; Uchino, A.; Nakajima, J.; Eriguchi, M.; Nanri, Y.; Hara, M.; Horikawa, E.; et al. Brain microbleeds and global cognitive function in adults without neurological disorder. *Stroke* **2008**, *39*, 3323–3328. [CrossRef] [PubMed]
35. van der Flier, W.M.; van Straaten, E.C.; Barkhof, F.; Verdelho, A.; Madureira, S.; Pantoni, L.; Inzitari, D.; Erkinjuntti, T.; Crisby, M.; Waldemar, G.; et al. Small vessel disease and general cognitive function in nondisabled elderly: The LADIS study. *Stroke* **2005**, *36*, 2116–2120. [CrossRef]
36. Hata, R.; Matsumoto, M.; Handa, N.; Terakawa, H.; Sugitani, Y.; Kamada, T. Effects of hemodialysis on cerebral circulation evaluated by transcranial Doppler ultrasonography. *Stroke* **1994**, *25*, 408–412. [CrossRef] [PubMed]
37. Kanai, H.; Hirakata, H.; Nakane, H.; Fujii, K.; Hirakata, E.; Ibayashi, S.; Kuwabara, Y. Depressed cerebral oxygen metabolism in patients with chronic renal failure: A positron emission tomography study. *Am. J. Kidney Dis.* **2001**, *38*, S129–S133. [CrossRef]
38. Murray, A.M. Cognitive impairment in the aging dialysis and chronic kidney disease populations: An occult burden. *Adv. Chronic Kidney Dis.* **2008**, *15*, 123–132. [CrossRef]
39. Longo, N.; Frigeni, M.; Pasquali, M. Carnitine transport and fatty acid oxidation. *Biochim. Biophys. Acta* **2016**, *1863*, 2422–2435. [CrossRef]
40. Varjacic, A.; Mantini, D.; Demeyere, N.; Gillebert, C.R. Neural signatures of Trail Making Test performance: Evidence from lesion-mapping and neuroimaging studies. *Neuropsychologia* **2018**, *115*, 78–87. [CrossRef] [PubMed]
41. Arbuthnott, K.; Frank, J. Trail making test, part B as a measure of executive control: Validation using a set-switching paradigm. *J. Clin. Exp. Neuropsychol.* **2000**, *22*, 518–528. [CrossRef]
42. Glascher, J.; Adolphs, R.; Damasio, H.; Bechara, A.; Rudrauf, D.; Calamia, M.; Paul, L.K.; Tranel, D. Lesion mapping of cognitive control and value-based decision making in the prefrontal cortex. *Proc. Natl. Acad. Sci. USA* **2012**, *109*, 14681–14686. [CrossRef] [PubMed]
43. Kopp, B.; Rosser, N.; Tabeling, S.; Sturenburg, H.J.; de Haan, B.; Karnath, H.O.; Wessel, K. Errors on the Trail Making Test Are Associated with Right Hemispheric Frontal Lobe Damage in Stroke Patients. *Behav. Neurol.* **2015**, *2015*, 309235. [CrossRef] [PubMed]
44. Miskin, N.; Thesen, T.; Barr, W.B.; Butler, T.; Wang, X.; Dugan, P.; Kuzniecky, R.; Doyle, W.; Devinsky, O.; Blackmon, K. Prefrontal lobe structural integrity and trail making test, part B: Converging findings from surface-based cortical thickness and voxel-based lesion symptom analyses. *Brain Imaging Behav.* **2016**, *10*, 675–685. [CrossRef] [PubMed]

Article

Nutrition in the First Week after Stroke Is Associated with Discharge to Home

Yoichi Sato [1], Yoshihiro Yoshimura [2,*] and Takafumi Abe [1]

1   Department of Rehabilitation, Uonuma Kikan Hospital, Minamiuonuma 949-7302, Japan; yoichi3041@gmail.com (Y.S.); t.abe.pt@gmail.com (T.A.)

2   Center for Sarcopenia and Malnutrition Research, Kumamoto Rehabilitation Hospital, Kumamoto 869-1106, Japan

*   Correspondence: hanley.belfus@gmail.com; Tel.: +81-96-232-3111

Abstract: Malnutrition is associated with poor clinical outcomes in stroke patients. The effect of early nutritional intake after admission on home discharge is unclear. We evaluated the impact of energy intake in the first week of hospitalization of acute stroke patients on home discharge and activities of daily living (ADL). A retrospective cohort study was conducted with 201 stroke patients admitted to an acute care hospital in Japan. The energy and protein intake during the first week were evaluated. Multivariate models were used to estimate variables related to discharge destination and ADL at discharge. The cut-off point of nutritional intake for determining the discharge destination was evaluated using the receiver operating characteristic curve. Out of 163 patients included in the analysis, 89 (54.6%) and 74 (45.4%) were discharged home and elsewhere, respectively. Those discharged home had higher energy and protein intake than those discharged elsewhere. In multiple regression analysis, energy intake was independently associated with ADL at discharge and home discharge (odds ratio 1.146). Those with energy intake >20.7 kcal/kg/day had higher ADL at discharge and more patients discharged home than those with energy intake <20.7 kcal/kg/day. Energy intake during the first week affected home discharge in acute stroke patients.

Keywords: energy intake; home-discharge; activity of daily living; stroke

Citation: Sato, Y.; Yoshimura, Y.; Abe, T. Nutrition in the First Week after Stroke Is Associated with Discharge to Home. *Nutrients* **2021**, *13*, 943. https://doi.org/10.3390/nu13030943

Academic Editor: Susanna C. Larsson

Received: 17 February 2021
Accepted: 12 March 2021
Published: 15 March 2021

Publisher's Note: MDPI stays neutral with regard to jurisdictional claims in published maps and institutional affiliations.

1. Introduction

Malnutrition in post-stroke patients results in increased mortality, complications, and poor functional prognosis [1–3]. Malnutrition is associated with dysphagia and impaired consciousness [3,4]. Prolonged starvation uses skeletal muscle for energy and causes muscle loss and muscle dysfunction [5], leading to the onset or exacerbation of sarcopenia, which is frequently observed in post-stroke patients, and stroke patients with sarcopenia have poor improvement in physical and swallowing functions [6,7].

Energy intake after admission is associated with functional improvement at discharge. In convalescent rehabilitation wards in Japan, energy intake during the first week after admission is associated with activities of daily living (ADL) at discharge in post-stroke patients [3]. In the acute phase, 26.4% of stroke patients are malnourished one week after admission, and high energy intake in the first week improves ADL at discharge [1,3].

Sex, stroke category, severity, level of ADL, and nutritional status at admission have been reported as factors predicting home discharge in acute stroke patients [8–11]. However, the effect of early nutritional intake after admission on home discharge is unclear. In addition, both energy and protein intake may affect the nutritional status of acute stroke patients [1]; however, the effect of protein intake on the discharge destination is unknown.

Protein-energy malnutrition worsens ADL after 30 days of admission in acute stroke patients [1] and may affect their discharge destination, because high ADL level is associated with home discharge [8,11]. Therefore, in this study, we evaluated the impact of energy and protein intake during the first week of admission on home discharge and ADL in acute stroke patients. The results suggest that early and active nutritional support in the acute phase may affect functional prognosis and post-discharge destination.

2. Materials and Methods

2.1. Participants and Setting

We conducted a single-center retrospective cohort study at a 454-bed acute care hospital in Niigata, Japan. We enrolled 201 consecutive patients who were admitted between May 2020 and January 2021 with cerebral infarction and cerebral hemorrhage within 48 h after onset. The presence of stroke was confirmed in all enrolled patients using computed tomography or magnetic resonance imaging [12]. The exclusion criteria were (1) missing data, (2) altered consciousness, (3) pacemaker implantation, (4) admission for diseases other than stroke, and (5) death. The observation period was the duration of hospitalization (from the date of admission to the date of discharge).

During the study period, 201 stroke patients were registered (Figure 1). Based on the exclusion criteria, 38 patients were excluded: missing data ($n = 18$), altered consciousness ($n = 10$), pacemaker implantation ($n = 3$), admission of diseases other than stroke ($n = 5$), and death during hospitalization ($n = 2$). Finally, 163 participants (mean age 75.2 ± 12.6 years; 36.8% women) were analyzed.

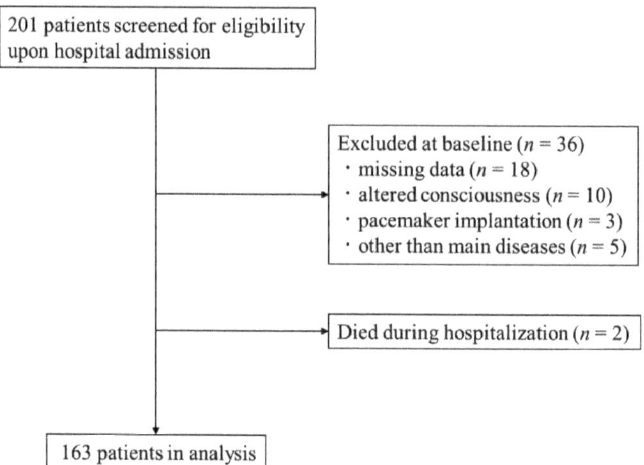

Figure 1. Flowchart of the study population.

The rehabilitation program (up to 3 h/day), aimed at improving endurance, ADL training, and dysphagia rehabilitation, was tailored to accommodate the functional abilities and disabilities of the individual patient and included paralyzed-limb facilitation, range-of-motion exercises, basic movement training (mainly for the legs), walking training, resistance training, and aerobic exercises using an ergometer [13,14]. Rehabilitation therapy was performed in a general way according to the patient's functional abilities and disabilities.

2.2. Data Collection

Basic information was recorded at admission, such as age, sex, body mass index (BMI), stroke type, stroke severity (based on National Institutes of Health Stroke Scale; NIHSS), days from onset, and other laboratory data (serum concentration of albumin and hemoglobin). Nutritional status was assessed using the Geriatric Nutritional Risk Index (GNRI) [15,16], which was calculated from the serum albumin concentration and body weight using the following equation: GNRI = [14.89 × albumin concentration (g/dL)] + [41.7 × (actual body weight/ideal body weight)]. Ideal body weight was defined as a BMI of 22.0 kg/m^2. Skeletal muscle mass and handgrip strength were measured within 5 days of admission. The grip strength was measured using a Smedley hand dynamometer, and the maximum value out of two measurements on each side was considered. In case of paralysis, the value of the hand without paralysis was considered. If both hands could not be measured due to coma, they were excluded from the handgrip strength results. Skeletal muscle mass was assessed using bioelectrical impedance analysis using InBody S10 (InBody, Tokyo, Japan), and the skeletal muscle index (SMI) was calculated based on skeletal muscle mass [17]. Swallowing function was assessed using the Functional Oral Intake Scale (FOIS) [18]. It was determined based on the dietary pattern after evaluation by a speech therapist within 3 days of admission.

2.3. Nutrition Intake

Based on a previous study [19], we reviewed diet records to quantify the mean daily nutritional intake during the first week after admission to the acute care hospital. Oral intake was measured by the experienced nurses using visual estimation. Visual estimation is commonly used in hospitals to evaluate food intake through the estimation of plate waste [14], and it comprises visually estimating the food present before and after the plate is provided to the patient. After the patient finished the meal, the nurse would register how much food the patient has taken in on an 11 point scale of 0–10. The oral nutritional intake was calculated from the amounts of calories and protein provided in the diet and amount of dietary intake. Energy and protein content during enteral and parenteral nutrition were collected from medical records. Energy and protein intake was defined as the mean value of nutritional intake in the first week divided by body weight.

2.4. Outcome Measurement

The primary outcome was the discharge destination from the acute care hospital, and this was categorized as home and others (convalescent rehabilitation hospital or nursing facility).

The secondary outcome was ADL, which was assessed using the Functional Independence Measure (FIM). The FIM score rates 13 motor and 5 cognitive activities on a scale of 1 (complete dependence) to 7 (complete independence). The total FIM score ranges from 18 to 126 points. A high FIM score indicates high activities of daily living [7]. The FIM gain was defined as the FIM at discharge minus the FIM at admission.

2.5. Sample Size Calculation and Statistical Analysis

We divided the patients into two groups: those discharged to home and those discharged to other facilities. To compare nutritional intake between the two groups, the null hypothesis was rejected with a sample size of at least 64 participants in each group with 0.5 effect size, 0.8 detection power, and 0.05 alpha error.

The results are reported as mean (standard deviation, SD) for parametric data, as median (25th to 75th percentile or interquartile range, IQR) for nonparametric data, and as number (%) for categorical data. The unpaired t test, Mann–Whitney U test, and Chi-square test were performed for comparison between the two groups. Multiple logistic regression analysis was used to determine whether nutritional intake was independently associated with home-discharge (primary outcome) from the acute care hospital. Based on previous studies [10,20–23], we selected age, sex, stroke category, NIHSS score, length of hospital stay, nutritional status, SMI, handgrip strength, swallowing function, FIM score at admission, FIM eating at discharge, FIM gain, paralysis (lower limbs), and rehabilitation time as covariates. If energy intake or protein intake had a significant effect on home discharge, the cut-off point of nutritional intake for determining the discharge destination was evaluated using the receiver operating characteristic (ROC) curve. We used the value at which the Youden Index was the highest as a criterion for determining the cut-off point. The patient characteristics were compared by dividing them into two groups at the cut-off point. Multiple linear regression analysis was used to determine whether nutritional intake was independently associated with FIM at discharge (secondary outcome). Using the same covariates (excluding FIM eating at discharge, FIM gain) as in the logistic regression analysis, we investigated the effects of energy and protein intake on ADL at discharge. Multicollinearity was assessed using the variance inflation factor (VIF): VIF value between 1 and 10 was considered as the absence of multicollinearity. All analyses were performed using SPSS version 21 (IBM, Armonk, NY, USA). $p < 0.05$ was considered statistically significant.

2.6. Ethics

This study was approved by the Institutional Review Board of the study center (02–024). Written informed consent could not be obtained because of the constraints imposed by the retrospective study design, although the participants could withdraw from this study at any time by using an opt-out procedure. This study was conducted in accordance with the Declaration of Helsinki and ethical guidelines for medical and health research involving human subjects.

3. Results

The baseline characteristics of the enrolled participants are summarized in Table 1. Of these, 89 (54.6%) and 74 (45.4%) were discharged home and elsewhere, respectively. The median FIM score at admission was 58 (IQR: 32–83), suggesting that a large number of patients were physically dependent at baseline. The NIHSS score at admission was 5 (2–9) points. The median length of hospital stay was 19 (11–28) days. The median rehabilitation time was 60.4 (45.8–76.0) min/day. The home discharge group had significantly younger patients ($p < 0.001$), more male patients ($p = 0.011$), and patients with lower severity of stroke, milder motor paralysis, and shorter length of hospital stay than the other discharge group (all $p < 0.001$). Moreover, SMI, handgrip strength, and swallowing function (FOIS) at admission were significantly higher in the home discharge group than the other discharge group (all $p < 0.001$). The rehabilitation time was significantly longer in the other discharge group than in the home discharge group ($p = 0.022$). The FIM score at admission in the other discharge group was 32 (20–51), and many patients required assistance for ADL. In contrast, the score in the home discharge group was 80 (66–93), and some patients showed early onset independence. The FIM eating at discharge was higher in the home discharge group ($p < 0.001$), but FIM gain was higher in the other discharge group ($p = 0.001$). The energy intake in the home discharge group was 23.5 (16.7–26.6) kcal/kg/day which was significantly higher than that in the other discharge group (12.4 (9.3–18.4) kcal/kg/day). Similarly, the protein intake in the home discharge group was 0.9 (0.8–1.1) g/kg/day, which was significantly higher than that in the other discharge group (0.7 (0.5–0.9) g/kg/day).

Table 2 shows the results of the multivariate analysis according to the discharge destination and FIM at discharge. There was no multicollinearity between the variables. Multiple logistic regression analysis showed that energy intake (odds ratio (OR) = 1.146, 95% confidence interval (CI) = 1.029–1.276, $p = 0.013$), FOIS (OR = 1.450, 95% CI = 1.036–2.544, $p = 0.036$), FIM at admission (OR = 1.039, 95% CI = 1.003–1.075, $p = 0.033$), and FIM eating at discharge (OR = 1.651, 95% CI = 1.952–2.063, $p = 0.045$) were significantly associated with home discharge. The Hosmer–Lemeshow test was $p = 0.887$ with high prediction accuracy. The discriminative predictive value of this logistic regression analysis was 87.5%. This result indicates that high energy intake is an independent predictor of home discharge. The multiple linear regression analysis shows that energy intake ($\beta = 0.131$, $p = 0.025$), length of stay ($\beta = 0.108$, $p = 0.041$), SMI ($\beta = 0.164$, $p = 0.019$), Handgrip strength ($\beta = 0.166$, $p = 0.028$), FIM at admission ($\beta = 0.349$, $p < 0.001$), and Brunnstrom recovery stage (BRS)-lower limb ($\beta = 0.123$, $p = 0.049$) were positively associated and NIHSS ($\beta = -0.164$, $p = 0.020$) was negatively and independently associated with FIM score at discharge (Adjusted R2 = 0.799). Protein intake was not a significant variable for home discharge or FIM score at discharge.

Table 1. Comparison of patient characteristics between the two groups by discharge destination.

	Total (*n* = 163)	Home (*n* = 89)	Other (*n* = 74)	*p*–Value
Age (year)	75.2 (12.6)	71.4 (12.6)	78.4 (11.8)	<0.001
Sex (male/female)	103/60	64/25	39/35	0.011
Body mass index (kg/m^2)	22.8 (3.9)	23.1 (3.9)	22.3 (3.9)	0.205
NIHSS score	5 (2–9)	2 (1–5)	9 (6–14)	<0.001
Stroke type (infarct/hemorrhage)	129/34	75/14	54/20	0.077
Comorbidity (%)				
Hypertension	109 (66.9)	61 (68.5)	48 (64.9)	0.547
Diabetes	54 (33.1)	26 (29.2)	28 (37.8)	0.244
Previous stroke	19 (11.7)	7 (7.9)	12 (16.2)	0.098
Atrial fibrillation	50 (30.7)	23 (25.8)	27 (36.5)	0.142
Side of lesion (right/left/both)	65/89/9	38/45/6	27/44/3	0.401
BRS				
Upper limb	5 (3–6)	6 (5–6)	3 (2–5)	<0.001
Hand-finger	5 (3–6)	5 (5–6)	3 (2–5)	<0.001
Lower limb	5 (4–6)	6 (5–6)	4 (2–5)	<0.001
Days from onset (day)	0 (0–0)	0 (0–0)	0 (0–0)	0.989
Length of hospital stay (day)	19 (11–28)	14 (10–19)	27 (20–34)	<0.001
Laboratory data				
Albumin (g/dL)	4.0 (0.5)	4.1 (0.5)	30.9 (00.5)	0.079
Hemoglobin (g/dL)	13.3 (2.1)	13.5 (2.1)	130.1 (20.2)	0.268
GNRI	103.3 (95.2–108.9)	104.7 (96.7–110.5)	101.0 (93.6–107.8)	0.275
SMI (kg/m^2)	7.0 (6.0–8.1)	7.2 (6.3–8.6)	6.4 (5.4–7.6)	<0.001
Handgrip strength (kg)	22.9 (14.0–30.2)	25.0 (18.5–33.0)	15.2 (8.6–20.0)	<0.001
FOIS	5 (2–6)	6 (5–7)	2 (1–5)	<0.001
Rehabilitation time (min/day)	60.4 (45.8–76.0)	54.8 (37.9–74.5)	64.0 (52.5–77.4)	0.022
FIM at admission	58 (32–83)	80 (66–93)	32 (20–51)	<0.001
FIM eating at discharge	7 (5–7)	7 (7–7)	4 (2–6)	<0.001
FIM gain	24 (11–40)	22 (11–40)	28 (11–39)	0.001
Energy intake (kcal/kg/day)	18.3 (12.1–24.7)	23.5 (16.7–26.6)	12.4 (9.3–18.4)	<0.001
Protein intake (g/kg/day)	0.9 (0.6–1.0)	0.9 (0.8–1.1)	0.7 (0.5–0.9)	<0.001

Mean (SD) or median (IQR) or subjects (%); BRS: Brunnstrom stage, GNRI: Geriatric Nutritional Risk Index, SMI: Skeletal Muscle Mass Index; FOIS: Functional Oral Intake Scale, FIM: Functional Independence Measure; Handgrip strength: Home (*n* = 78), Other (*n* = 42).

The ROC curve of energy intake for the determination of home discharge is shown in Figure 2. The area under the curve (95% CI) for the energy intake was 0.795 (0.726–0.864). Energy intake was a significant predictive variable ($p < 0.001$). The cut-off point of energy intake for determining home discharge was 20.7 kcal/kg/day (sensitivity, 0.618; specificity, 0.838). Table 3 shows the evaluation of patients in each group divided according to the cut-off point: those above the cut-off point (High Group) and those below the cut-off point (Low Group). There was no significant difference in the GNRI between the two groups. The length of hospital stay was significantly shorter in the High Group than in the Low Group ($p = 0.012$). The FOIS and FIM scores at discharge and home discharge rate were significantly higher in the High Group than in the Low Group (all $p < 0.001$).

Table 2. Multivariate analysis for home discharge and FIM at discharge.

	Home–Discharge [#1]		FIM at Discharge [#2]	
	OR (95% CI)	*p*–Value	β	*p*-Value
Age	0.966 (0.903–1.034)	0.323	−0.065	0.131
Gender (male)	1.160 (0.214–6.293)	0.864	−0.048	0.302
Stroke type(infarction)	0.466 (0.079–2.729)	0.397	−0.084	0.109
NIHSS	0.836 (0.637–1.098)	0.198	−0.164	0.020
Length of stay	0.974 (0.937–1.012)	0.180	0.108	0.041
GNRI	1.012 (0.977–1.048)	0.514	0.060	0.205
SMI	1.137 (0.551–2.345)	0.729	0.164	0.019
Handgrip strength	1.057 (0.922–1.212)	0.425	0.166	0.028
FOIS	1.450 (1.036–2.544)	0.036	0.070	0.278
FIM at admission	1.039 (1.003–1.075)	0.033	0.349	<0.001
FIM eating at discharge	1.651 (1.952–2.063)	0.045		
FIM gain	1.042 (0.979–1.109)	0.193		
BRS-lower limb	0.819 (0.432–1.549)	0.539	0.123	0.049
Rehabilitation therapy	0.968 (0.941–1.083)	0.074	0.109	0.121
Energy intake	1.146 (1.029–1.276)	0.013	0.131	0.025
Protein intake	0.104 (0.006–1.739)	0.115	0.063	0.264

[#1] Multiple logistic regression analysis, [#2] Multiple linear regression analysis; NIHSS: National Institutes of Health Stroke Scale, GNRI: Geriatric Nutritional Risk Index; SMI: Skeletal Muscle Mass Index, FOIS: Functional Oral Intake Scale; FIM: Functional Independence Measure, BRS: Brunnstrom stage.

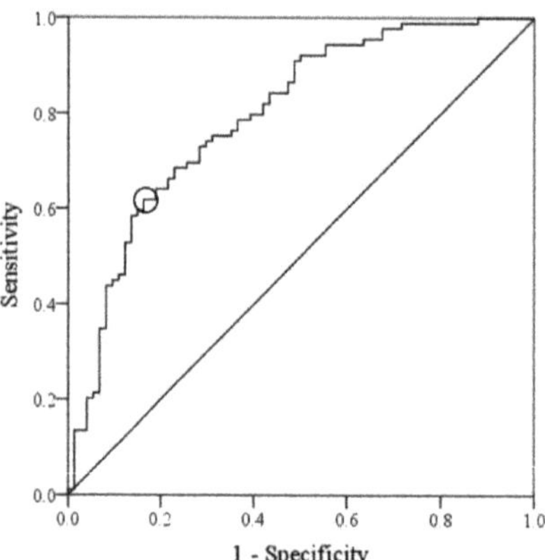

Figure 2. Receiver-operating characteristic curve analysis for home-discharge. The cut-off points of energy intake for determining home discharge rate were 20.7 kcal/kg/day (sensitivity 0.618, specificity 0.838, area under the curve 0.795, 95% confidence interval 0.726–0.864, $p < 0.001$); the circle indicates cut-off point.

Table 3. Comparison between two groups divided by cut-off point of energy intake.

	High Group (*n* = 67)	Low Group (*n* = 96)	*p*-Value
Length of hospital stay (day)	14 (10–24)	22 (16–31)	0.012
GNRI	103.3 (94.7–108.3)	102.9 (95.1–109.4)	0.270
FOIS	6 (5–7)	4 (1–5)	<0.001
FIM at discharge	122 (97–126)	78 (40–111)	<0.001
Discharge to home (%)	55 (82.1)	34 (35.4)	<0.001

Median (IQR) or subjects (%); GNRI: Geriatric Nutritional Risk Index, FOIS: Functional Oral Intake Scale; FIM: Functional Independence Measure; two groups divided by cut-off point of energy intake (20.7 kcal/kg/day).

4. Discussion

In this study, we reported the relationship between nutritional intake and discharge destination in the early phase of hospitalization in acute stroke patients. We report that in acute stroke patients: (1) energy intake during the first week of admission was independently associated with home discharge; (2) 20.7 kcal/kg/day might be the cut-off point for predicting home discharge; (3) protein intake was not related to home discharge and the level of ADL at discharge.

Energy intake in the early phase after onset in acute stroke patients is associated with functional prognosis. In a previous study on acute stroke patients, high energy intake in the early phase after onset resulted in significant improvements in ADL [24]. Kokura et al. measured energy intake for one week after admission and compared it with the basal energy expenditure calculated using the Harris–Benedict equation [19]. They reported that the energy-sufficient group had a higher FIM gain at discharge than the energy-deficient group, and many patients had improved nutritional status. However, only 39.1% of patients were energy-sufficient, indicating that many acute stroke patients had inadequate energy intake [19]. Malnutrition is present in 26.4% and 35% of acute stroke patients in the first and second week after admission, respectively [1]. In this study, we did not assess body weight at discharge, so we did not know how many patients lost weight due to energy deficient. Aggressive nutritional support in the early phase after onset, when there is a high risk of malnutrition, may lead to significant improvements in ADL. Moreover, it has been reported that a high level of ADL was associated with high home discharge rate [10,11,21,25]. This evidence supports our findings. Furthermore, consistent with previous studies, this study reports that good swallowing function was independently associated with home discharge [8,9]. Therefore, early detection of dysphagia is essential in the nutritional management of post-stroke patients, including dietary modification.

Energy intake predicting home discharge was 20.7 kcal/kg/day in this study population. High energy intake patients above this cut-off point exhibited higher ADL recovery and home discharge rates than those below the cut-off point. In a previous study on acute stroke patients, the energy sufficient group (median: 21.5 kcal/kg/day) had a higher ADL recovery than the deficient group (median: 10.3 kcal/kg/day) [18]. Acute stroke patients with inadequate energy intake (median: 11.6 kcal/kg/day) reported more loss of quadriceps muscle on the non-paralyzed side than those with adequate energy intake (median: 24.0 kcal/kg/day) [19]. Therefore, the cut-off point of this study supports the findings of previous studies. Contrarily, in convalescent rehabilitation hospitals, post-stroke patients with severe malnutrition (GNRI < 82) had an average energy intake of 26.9 kcal/kg/day [16]. However, the study reported that this energy intake may not be sufficient to improve weight loss in patients with severe malnutrition [16]. Thus, it is necessary to provide appropriate energy intake depending on the phase of the disease and the status of malnutrition.

In this study, protein intake did not affect the discharge destination or FIM score at discharge. There are two possible reasons for this. First, the participants of this study included patients who were provided with a protein-restricted diet due to renal dysfunction. In the elderly, 1.0–1.2 g/kg/day of protein intake is recommended to maintain or improve lean body mass [26,27]. However, it is possible that some patients were provided with a diet with less protein content than recommended due to renal dysfunction. Second, protein quality may affect protein intake. Randomized controlled trials in Japan showed that the combination of "leucine-rich" or "branched-chain amino acid intake" and "resistance training" improved muscle mass, strength, and physical function in stroke patients with sarcopenia [28–30]. Therefore, the intake of high-quality amino acids, but not total protein content, may predict the discharge destination and functional prognosis of stroke patients.

The present study has several limitations. First, because it was a retrospective cohort study in a single acute care hospital, we cannot generalize the results, eliminate potential confounders, or prove a causal relationship between energy intake and discharge destination. Second, nurses visually assessed nutritional intake, the main parameter, which may have resulted in measurement errors. Third, the physical functions of patients with impaired consciousness could not be accurately assessed. For example, the handgrip strength of coma patients was excluded. Fourth, we did not record the foods that family members brought from the outside. It is possible that they are taking in more nutrients than we have assessed. These limitations can be overcome by conducting a prospective cohort study at multiple centers. In the future, it will be necessary to examine the relationship between energy intake and discharge destination in interventional studies.

5. Conclusions

Energy intake during the first week of admission is associated with discharge destination and ADL at discharge in acute stroke patients. This finding suggests that early detection of nutrition-related factors, such as dysphagia and aggressive nutritional support, may enhance home discharge in this setting.

Author Contributions: Conceptualization, Y.S. and T.A.; data curation, Y.S. and T.A.; formal analysis, Y.S.; writing—original draft preparation; and writing—review and editing, Y.S. and Y.Y. All authors have read and agreed to the published version of the manuscript.

Funding: This research received no external funding.

Institutional Review Board Statement: This study was approved by the Institutional Review Board of the study center. This study was conducted in accordance with the Declaration of Helsinki and ethical guidelines for medical and health research involving human subjects.

Informed Consent Statement: Patient consent was waived due to the constraints imposed by the retrospective study design, although the participants could withdraw from this study at any time by using an opt-out procedure.

Data Availability Statement: The data are not publicly available owing to opt out restrictions. Data sharing is not applicable.

Acknowledgments: We greatly appreciate the staff at Uonuma Kikan Hospital who supported this study.

Conflicts of Interest: The authors declare no conflict of interest.

References

1. Dávalos, A.; Ricart, W.; Gonzalez-Huix, F.; Soler, S.; Marrugat, J.; Molins, A.; Suñer, R.; Genís, D. Effect of malnutrition after acute stroke on clinical outcome. *Stroke* **1996**, *27*, 1028–1032. [CrossRef]
2. Collaboration, F.T. Poor nutritional status on admission predicts poor outcomes after stroke: Observational data from the FOOD trial. *Stroke* **2003**, *34*, 1450–1456. [CrossRef] [PubMed]
3. Nii, M.; Maeda, K.; Wakabayashi, H.; Nishioka, S.; Tanaka, A. Nutritional improvement and energy intake are associated with functional recovery in patients after cerebrovascular disorders. *J. Stroke Cerebrovasc. Dis.* **2016**, *25*, 57–62. [CrossRef]
4. Chen, N.; Li, Y.; Fang, J.; Lu, Q.; He, L. Risk factors for malnutrition in stroke patients: A meta-analysis. *Clin. Nutr.* **2019**, *38*, 127–135. [CrossRef] [PubMed]
5. Pichard, C.; Jeejeebhoy, K.N. Muscle dysfunction in malnourished patients. *Q. J. Med.* **1988**, *69*, 1021–1045.
6. Shiraishi, A.; Yoshimura, Y.; Wakabayashi, H.; Tsuji, Y. Prevalence of stroke-related sarcopenia and its association with poor oral status in post-acute stroke patients: Implications for oral sarcopenia. *Clin. Nutr.* **2018**, *37*, 204–207. [CrossRef] [PubMed]
7. Yoshimura, Y.; Wakabayashi, H.; Bise, T.; Tanoue, M. Prevalence of sarcopenia and its association with activities of daily living and dysphagia in convalescent rehabilitation ward inpatients. *Clin. Nutr.* **2018**, *37*, 2022–2028. [CrossRef] [PubMed]
8. Nguyen, V.Q.; PrvuBettger, J.; Guerrier, T.; Hirsch, M.A.; Thomas, J.G.; Pugh, T.M.; Rhoads, C.F., 3rd. Factors associated with discharge to home versus discharge to institutional care after inpatient stroke rehabilitation. *Arch. Phys. Med. Rehabil.* **2015**, *96*, 1297–1303. [CrossRef] [PubMed]
9. Kenmuir, C.L.; Hammer, M.; Jovin, T.; Reddy, V.; Wechsler, L.; Jadhav, A. Predictors of outcome in patients presenting with acute ischemic stroke and mild stroke scale scores. *J. Stroke Cerebrovasc. Dis.* **2015**, *24*, 1685–1689. [CrossRef]
10. Suzuki, Y.; Tsubakino, S.; Fujii, H. Eating and grooming abilities predict outcomes in patients with early middle cerebral infarction: A retrospective cohort study. *Occup. Ther. Int.* **2020**, *2020*, 1374527. [CrossRef]
11. Sato, M.; Ido, Y.; Yoshimura, Y.; Mutai, H. Relationship of malnutrition during hospitalization with functional recovery and postdischarge destination in elderly stroke patients. *J. Stroke Cerebrovasc. Dis.* **2019**, *28*, 1866–1872. [CrossRef]
12. Nozoe, M.; Kanai, M.; Kubo, H.; Yamamoto, M.; Shimada, S.; Mase, K. Prestroke sarcopenia and functional outcomes in elderly patients who have had an acute stroke: A prospective cohort study. *Nutrition* **2019**, *66*, 44–47. [CrossRef]

13. Yoshimura, Y.; Wakabayashi, H.; Nagano, F.; Bise, T.; Shimazu, S.; Shiraishi, A. Chair-stand exercise improves post-stroke dysphagia. *Geriatr. Gerontol. Int.* **2020**, *20*, 885–891. [CrossRef]

14. Shimazu, S.; Yoshimura, Y.; Kudo, M.; Nagano, F.; Bise, T.; Shiraishi, A.; Sunahara, T. Frequent and personalized nutritional support leads to improved nutritional status, activities of daily living, and dysphagia after stroke. *Nutrition* **2020**, *83*, 111091. [CrossRef] [PubMed]

15. Kokura, Y.; Maeda, K.; Wakabayashi, H.; Nishioka, S.; Higashi, S. High nutritional-related risk on admission predicts less improvement of functional independence measure in geriatric stroke patients: A retrospective cohort study. *J. Stroke Cerebrovasc. Dis.* **2016**, *25*, 1335–1341. [CrossRef]

16. Nishioka, S.; Okamoto, T.; Takayama, M.; Urushihara, M.; Watanabe, M.; Kiriya, Y.; Shintani, K.; Nakagomi, H.; Kageyama, N. Malnutrition risk predicts recovery of full oral intake among older adult stroke patients undergoing enteral nutrition: Secondary analysis of a multicentre survey (the APPLE study). *Clin. Nutr.* **2017**, *36*, 1089–1096. [CrossRef]

17. Abe, T.; Iwata, K.; Yoshimura, Y.; Shinoda, T.; Inagaki, Y.; Ohya, S.; Yamada, K.; Oyanagi, K.; Maekawa, Y.; Honda, A.; et al. Low muscle mass is associated with walking function in patients with acute ischemic stroke. *J. Stroke Cerebrovasc. Dis.* **2020**, *29*, 105259. [CrossRef] [PubMed]

18. Kokura, Y.; Wakabayashi, H.; Nishioka, S.; Maeda, K. Nutritional intake is associated with activities of daily living and complications in older inpatients with stroke. *Geriatr. Gerontol. Int.* **2018**, *18*, 1334–1339. [CrossRef]

19. Kokura, Y.; Kato, M.; Taniguchi, Y.; Kimoto, K.; Okada, Y. Energy intake during the acute phase and changes in femoral muscle thickness in older hemiplegic inpatients with stroke. *Nutrition* **2020**, *70*, 110582. [CrossRef]

20. Yoshimura, Y.; Wakabayashi, H.; Nagano, F.; Bise, T.; Shimazu, S.; Kudo, M.; Shiraishi, A. Sarcopenic obesity is associated with activities of daily living and home discharge in post-acute rehabilitation. *J. Am. Med. Dir. Assoc.* **2020**, *21*, 1475–1480. [CrossRef] [PubMed]

21. Chung, D.M.; Niewczyk, P.; DiVita, M.; Markello, S.; Granger, C. Predictors of discharge to acute care after inpatient rehabilitation in severely affected stroke patients. *Arch. Phys. Med. Rehabil.* **2012**, *91*, 387–392.

22. Yoshimura, Y.; Bise, T.; Nagano, F.; Shimazu, S.; Shiraishi, A.; Yamaga, M.; Koga, H. Systemic Inflammation in the Recovery Stage of Stroke: Its Association with Sarcopenia and Poor Functional Rehabilitation Outcomes. *Prog. Rehabil. Med.* **2018**, *3*, 20180011. [CrossRef] [PubMed]

23. Yoshimura, Y.; Wakabayashi, H.; Bise, T.; Nagano, F.; Shimazu, S.; Shiraishi, A.; Yamaga, M.; Koga, H. Sarcopenia is associated with worse recovery of physical function and dysphagia and a lower rate of home discharge in Japanese hospitalized adults undergoing convalescent rehabilitation. *Nutrition* **2019**, *61*, 111–118. [CrossRef] [PubMed]

24. Nip, W.F.; Perry, L.; McLaren, S.; Mackenzie, A. Dietary intake, nutritional status and rehabilitation outcomes of stroke patients in hospital. *J. Hum. Nutr. Diet* **2011**, *24*, 460–469. [CrossRef] [PubMed]

25. Vluggen, T.; van Haastregt, J.C.M.; Tan, F.E.S.; Kempen, G.; Schols, J.; Verbunt, J.A. Factors associated with successful home discharge after inpatient rehabilitation in frail older stroke patients. *BMC Geriatr.* **2020**, *20*, 25. [CrossRef] [PubMed]

26. Bauer, J.; Biolo, G.; Cederholm, T.; Cesari, M.; Cruz-Jentoft, A.J.; Morley, J.E.; Phillips, S.; Sieber, C.; Stehle, P.; Teta, D.; et al. Evidence-based recommendations for optimal dietary protein intake in older people: A position paper from the PROT-AGE Study Group. *J. Am. Med. Dir. Assoc.* **2013**, *14*, 542–559. [CrossRef] [PubMed]

27. Yoshimura, Y.; Wakabayashi, H.; Yamada, M.; Kim, H.; Harada, A.; Arai, H. Interventions for Treating Sarcopenia: A Systematic Review and Meta-Analysis of Randomized Controlled Studies. *J. Am. Med. Dir. Assoc.* **2017**, *18*, 553.e1–533.e16. [CrossRef]

28. Yoshimura, Y.; Uchida, K.; Jeong, S.; Yamaga, M. Effects of nutritional supplements on muscle mass and activities of daily living in elderly rehabilitation patients with decreased muscle mass: A randomized controlled trial. *J. Nutr. Health Aging* **2016**, *20*, 185–191. [CrossRef]

29. Takeuchi, I.; Yoshimura, Y.; Shimazu, S.; Jeong, S.; Yamaga, M.; Koga, H. Effects of branched-chain amino acids and vitamin D supplementation on physical function, muscle mass and strength, and nutritional status in sarcopenic older adults undergoing hospital-based rehabilitation: A multicenter randomized controlled trial. *Geriatr. Gerontol. Int.* **2019**, *19*, 12–17. [CrossRef]

30. Yoshimura, Y.; Bise, T.; Shimazu, S.; Tanoue, M.; Tomioka, Y.; Araki, M.; Nishino, T.; Kuzuhara, A.; Takatsuki, F. Effects of a leucine-enriched amino acid supplement on muscle mass, muscle strength, and physical function in post-stroke patients with sarcopenia: A randomized controlled trial. *Nutrition* **2019**, *58*, 1–6. [CrossRef]

MDPI
St. Alban-Anlage 66
4052 Basel
Switzerland
Tel. +41 61 683 77 34
Fax +41 61 302 89 18
www.mdpi.com

Nutrients Editorial Office
E-mail: nutrients@mdpi.com
www.mdpi.com/journal/nutrients